An Introduction to Biomedical Science in Professional and Clinical Practice

An Introduction to Biomedical Science in Professional and Clinical Practice

by

Sarah J. Pitt
University of Brighton and
Brighton and Sussex University Hospitals NHS Trust, Brighton, UK

and

James M. Cunningham
University of Brighton, Brighton, UK

A John Wiley & Sons, Ltd., Publication

This edition first published 2009
© 2009 by John Wiley & Sons, Ltd.

Wiley-Blackwell is an imprint of John Wiley & Sons, formed by the merger of Wiley's global Scientific, Technical and Medical business with Blackwell Publishing.

Registered office: John Wiley & Sons, Ltd, The Atrium, Southern Gate, Chichester, West Sussex PO19 8SQ, UK

Other Editorial Offices:
9600 Garsington Road, Oxford OX4 2DQ, UK
111 River Street, Hoboken, NJ 07030-5774, USA

For details of our global editorial offices, for customer services and for information about how to apply for permission to reuse the copyright material in this book please see our website at www.wiley.com/wiley-blackwell

Library of Congress Cataloguing-in-Publication Data

Pitt, Sarah J.
    An introduction to biomedical science in professional and clinical practice / by Sarah J. Pitt
    and James M. Cunningham.
        p.   ;   cm.
    Includes bibliographical references and index.
    ISBN 978-0-470-05714-8 (hb) — ISBN 978-0-470-05715-5 (pbk.)
    1. Pathologists—Vocational guidance.   2. Medical scientists—Vocational guidance.
    I. Cunningham, James M.   II. Title.
    [DNLM: 1. Pathology, Clinical—methods.   2. Professional Competence.   3. Pathology,
    Clinical—organization & administration.   QY 21 P688i 2009]
    RB37.6.P58 2009
    616.07023—dc22

                                                                                                2008044203

A catalogue record for this book is available from the British Library.

ISBN: 978-0-470-05714-8 (H/B)
ISBN: 978-0-470-05715-5 (P/B)

Set in 11/13pt Times by Integra Software Services Pvt. Ltd, Pondicherry, India
Printed in Great Britain by CPI Antony Rowe, Chippenham, Wiltshire.

First Impression   2009

SP: For my husband Alan and my parents, John and Margaret
JC: For my partner Lesley and my family

Contents

Preface

This book is intended to help biomedical scientists undertaking their pre-registration training; we aim to help them understand the important aspects of working as a professional biomedical scientist in the clinical laboratory. It will be useful for students on 'coterminous' or 'integrated' Applied Biomedical Science BSc programmes, graduate trainee biomedical scientists and other staff who need to be aware of the knowledge and competency required to be a Health Professions Council registered practitioner. The book will also help qualified staff involved in training, by providing background information, key definitions and sources of further information. It brings together the essential information into a single text and points the reader towards more detailed sources where they might be appropriate. The suggested exercises that accompany each chapter are designed to provide ideas for suitable evidence for the Institute of Biomedical Science Registration Portfolio; the trainee and trainer can work together to adapt them to the local situation.

The recently established Applied Biomedical Science degrees in biomedical science, which this book is intended to support, arose as ventures between universities and clinical pathology departments across the UK. These programmes, supported by the professional body, have in our view given rise to new levels of excellence in partnership between academics and practitioners and are proven to be 'fit for purpose' by providing graduates who are 'fit to practise' by meeting the standards for professional registration. In addition to being a resource to support student-trainees and educator-trainers, we hope that this book exemplifies the success that this closer collaboration between academics and practitioners in biomedical science has brought. As the nature of the professional skills practised by biomedical scientists evolves in response to innovations in diagnostic science and changing patterns of healthcare delivery, this partnership should be an invaluable foundation to support and shape further the future developments in healthcare science education and training.

Acknowledgements

We would like to thank our colleagues at Brighton and Sussex University Hospitals NHS Trust and the University of Brighton for their expert advice and practical assistance, especially Sarah Bastow, Lindsey Dixon, Natasha Maderson, Anne Trezise, Helen Varney, Gary Weaving and Lakshmi Yaliwal. We are also grateful to Rebecca, Luke and Katherine Hulatt, Christopher Dark, Georgina Ellis, Ewan Lomax, Lynnsay Fuller, Sylvia Meisenberg, Andrew Parr and Anna Tarasewicz for contributing to the photographs. Particular thanks are due to Dr Alan Gunn for his support by providing suggestions for the text and supplying some of the figures, along with tireless proof-reading. We would also like to thank Rachael Ballard, Fiona Woods and their colleagues at John Wiley and Sons for their help and guidance.

1

Introduction to a career as a biomedical scientist

1.1 What is a biomedical scientist?

Biomedical scientists are scientifically qualified, registered practitioners who work in clinical pathology departments. They play a vital role in patient care, by carrying out diagnostic tests on samples such as blood, tissue and urine. As healthcare professionals, biomedical scientists work with a range of staff in hospitals and in primary care, to provide clinical laboratory services. About 70% of clinical diagnoses rely on pathology test results. This means that the laboratory work must be performed to the highest possible standards, that the correct specimen type from the right patient must be tested and that the results must be available, to the staff treating the patient, in good time.

Pathology test results are used to:

- **Diagnose** illnesses. For example, a person with symptoms of tiredness and dizziness might be suffering from iron deficiency anaemia, which could be confirmed by a blood test for haemoglobin concentration.

- **Monitor** conditions. For example, blood HbA1c levels are regularly monitored in diabetic patients to check that their blood sugar levels are well controlled.

- **Screen** for diseases in people who are at risk of having the condition, but do not appear to be ill, such as the test for *Chlamydia* infection.

The laboratory testing is performed or supervised by registered biomedical scientists (Figure 1.1), so although they do not have as much direct contact with patients as doctors and nurses, their contribution is vital. If there were not enough nurses available in a particular hospital, individual wards might have to be closed, whereas with insufficient biomedical scientists to run the pathology service, the work of the whole hospital would be compromised. Since the main role of biomedical scientists is to choose the most suitable testing

An Introduction to Biomedical Science in Professional and Clinical Practice   Sarah J. Pitt and James M. Cunningham
© 2009 John Wiley & Sons, Ltd

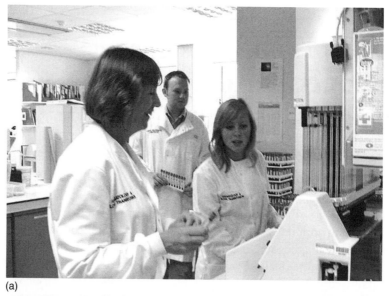

(a)

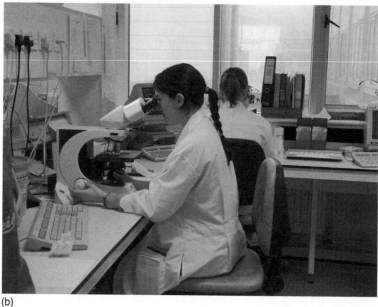

(b)

Figure 1.1   (a) Biomedical scientists discuss the operation of a laboratory analyser. (b) Biomedical scientists examine slides prepared from patient samples under a light microscope

method and then carry out the laboratory work accurately, it is clearly important that they are well trained and work to the highest standards at all times. This means that after the initial training, biomedical scientists have to continue to learn new techniques and keep up to date with scientific advances.

A career as a biomedical scientist can be very rewarding, as practitioners use scientific knowledge and technological expertise to help in diagnosis of disease and prevention of illness. They usually specialize in one clinical area of pathology: cytology, clinical chemistry, haematology, histopathology, immunology, medical microbiology, transfusion science or virology, although some posts cover more than one of these areas. In some pathology departments, biomedical scientists are qualified to carry out tests covering several of these areas, such as chemistry and haematology, where the same analyser can be used for all tests (see Chapter 6). Members of staff who are managers, such as quality managers, training managers and overall pathology service managers, need to be familiar with all disciplines, even if they originally trained and qualified in one area. For those who want to progress, there are many possibilities for career development within biomedical science. Some practitioners specialize in a particular technique or disease, through advanced training and scientific research. Others take on roles outside the traditional laboratory setting such as 'near patient testing' (see Chapter 2), lecturing at a university or working as a representative for a company which makes laboratory equipment and reagents. Biomedical scientists can also develop expertise as managers, which takes them to senior roles within pathology or the wider health service (see below and Chapter 2).

To qualify as a biomedical scientist, both academic and vocational training are required. The academic part of the qualification in the UK is usually a BSc in Biomedical Science which has been accredited by the Institute of Biomedical Science (see below); the vocational training must be undertaken in a approved training laboratory and is designed to show that the trainee has met the Standards of Proficiency set by the Health Professions Council (see below). It is possible to gain an accredited BSc and then take a post as a trainee Biomedical Scientist to complete the vocational training. However, a number of universities across the UK now offer courses which incorporate the vocational training – the so-called 'coterminous' or 'integrated' Applied Biomedical Science degree programmes.

1.2 Early development of clinical laboratory sciences

The idea of examining patients' specimens to aid the diagnosis of clinical disease was used centuries before the underlying science was understood. For example, the ancient literature (from around 1500 BC onwards) includes descriptions of the condition that we now recognize as diabetes mellitus. These relate symptoms of weight loss, extreme thirst and frequent micturation and describe sweet-tasting urine, which implies that someone would test the urine by drinking a little! This seems strange now that we have chemical tests for sugar in the urine (glycosuria) and can observe a colour change rather than a sugary taste, but it is an early example of a how a 'scientific' test can confirm a clinical diagnosis when the symptoms are fairly non-specific. Another example is an account from ancient Greek literature by Hippocrates of the progression of the respiratory infection known as 'consumption' (i.e. tuberculosis), which describes the patient's sputum as typically thick, greenish and sweet. A test involving the patient spitting on hot coals was used to determine the prognosis; a characteristic heavy odour noticed in the sputum, when it was heated on the coals, was taken as a sign that the patient would not survive.

From the 16th century onwards, scientific understanding deepened and broadened, as many discoveries were made and laboratory equipment such as microscopes and glassware was refined. This allowed the invention of more specific and reliable clinical diagnostic tests. By the 19th century, with an understanding of chemistry and the knowledge that the 'sweet taste' in urine is due to the presence of glucose, it was possible to devise a test for glycosuria. It involved adding a reagent containing copper sulfate to the sample and heating. The sugar acts as a reducing agent, changing the copper sulfate to copper oxide, which causes the solution to change colour from blue to reddish brown. Similarly, by the 1880s, once it had been established that infections were caused by microorganisms, light microscopes were powerful enough to detect them and stains were used to highlight them in samples. In this way, Robert Koch was able to identify a bacterium in the sputum samples of patients with 'consumption', which he named *Mycobacterium tuberculosis*. Both the Benedict's test for sugars in urine and the Ziehl – Neelson stain for tubercle bacilli are still used in clinical laboratories in many parts of the world today.

By the turn of the 20th century, as the benefits of science to medical diagnosis had become clearer, doctors were increasingly doing tests on patient's samples. These were usually carried out either by the bedside or in a room nearby that contained the necessary equipment and reagents. The potential for the results of such tests to be used to enhance patient management was recognized and doctors with an interest in laboratory diagnosis started to expand their repertoire. Although there were still some scientists and doctors researching in isolation, there was a demand for diagnostic services from colleagues within hospitals and 'routine' testing of samples began in conjunction with research. The work load therefore increased and hospitals began to employ 'laboratory assistants', who could collect and examine samples. These laboratory assistants were not medically qualified, but had a high level of scientific and technical training and their work contributed to improvements in accuracy and reliability of testing methods. However, as 'laboratory assistant' was not a recognized trade or profession, these workers did not have the protection of a trade union; hospital administrators considered them to be 'unskilled' and paid them low wages. Despite this, there are many examples of intelligent and motivated people who made significant, but unheralded, contributions to medical laboratory science at this time. A small selection of the important developments in medical science which affected diagnostic pathology are given in Box 1.1, but there are many other examples.

Box 1.1   Some examples of significant developments in clinical diagnostic pathology

- Development of the Haematoxylin and Eosin stain by Wissowsky in 1876

- Introduction of agar as a bacterial culture medium and use of covered dishes by Koch and Petri in the 1880s

- Invention of the Gram stain by Gram in 1884

- Characterization of the ABO blood group system by Landsteiner in 1902

- Invention of the radioimmunoassay by Yalow and Berson in 1960

- Development of the technique to manufacture monoclonal antibodies by Milstein and Köhler in 1975

- Invention of the Polymerase Chain Reaction by Mullis in 1983

1.3   Development of the biomedical science profession

By the early 20th century, many people were working to provide technical and scientific support for clinical diagnosis in laboratories all over the UK. However, the opportunities for staff working in different laboratories to communicate experiences and ideas were limited and there was a clear need for a professional organization to facilitate this. A laboratory scientist working in Liverpool, Albert Norman, recognized the requirement for such an organization to support his colleagues. Norman consulted with medical colleagues who had formed the Pathological Society of Great Britain and Northern Ireland in 1906, and so had recent experience of founding and running a professional organization. He always believed that the two professional groups should cooperate closely for the good of patient care. Norman founded the Pathological and Bacteriological Laboratory Assistants' Association (PBLAA) in 1912. The name reflected the nature of the work that laboratory assistants were undertaking at that time. Stains had been developed for use on tissue samples, which allowed the discrimination of cell types and identification of abnormalities when they were examined under the microscope. Agar-based media for the culture of bacteria were also available by this time, which meant that organisms could be grown from patients' samples and stains were used to help visualize pathogens under the microscope. Some basic biochemical tests were also performed.

In order to be a full member of the PBLAA, a laboratory assistant needed to have undertaken 3 years of training and pay a subscription of 5 shillings (which is 25p in decimal currency but was a considerable amount of money at the time!). Through its members, the PBLAA organized scientific meetings, at which findings from research and developments in diagnostic testing techniques could be shared and also social events. Meeting colleagues under the auspices of the professional body, to discuss scientific issues or just to enjoy each other's company, is still important for biomedical scientists today. In 1913, the PBLAA started publication of the *Laboratory Journal*, which was the predecessor of the *British Journal of Biomedical Science*. Then, as now, the purpose was to provide a journal in which members could publish articles describing their scientific research and improvements in techniques to share with each other. This was even more important in 1913 than it is now, as people did not travel as much as they do in the 21st century for work-related purposes, few people had access to telephones and it would be another 80 years before the Internet was invented!

During the First World War, PBLAA members were often conscripted to the front line rather than to work in field hospitals, because they had no formal professional qualifications. This was obviously a waste of their talents and abilities and so after the War, it was decided to introduce written and practical examinations for laboratory assistants. The first examinations were held in 1921 and covered pathological and bacteriological techniques. The idea was to give structure to training and to ensure that laboratory assistants in all clinical laboratories in the country were operating to the same high standards. In 1937, the professional examination comprised two stages, Part I and Part II. A laboratory assistant who passed Part I was eligible to become an 'Associate' of the PBLAA, while the attainment of Part II was necessary to be a 'Member' and was a prerequisite for employment in senior posts. A more senior designation of 'Fellow' was also introduced. In time, advances in pathological sciences led to people concentrating on haematological investigations, blood transfusion or biochemical tests, to the extent that these became recognized as separate disciplines; examinations in these particular subjects were introduced to reflect this and to support the career development of members working in these areas. During the latter half of the 20th century, further discoveries led to the setting up of more specialized sections within pathology, including immunology, virology and cytology and examinations are now offered in these areas also.

In 1943, the PBLAA changed its name to the Institute of Medical Laboratory Technology (IMLT) and its members were known as 'medical laboratory technicians' (MLTs). Professional status for UK members was eventually consolidated by the passing of the Professions Supplementary to Medicine Act in 1960; this made it a legal requirement that MLTs must be registered with the Council for Professions Supplementary to Medicine (CPSM) before practising in the National Health Service. However, by the 1970s, the term 'technician' seemed outdated and not a fair reflection of the high levels of training and qualification required to do the job. Members of the profession were involved in research and development in addition to routine diagnostic work and they wanted a title which acknowledged the scientific nature of their work. The IMLT was renamed the Institute of Medical Laboratory Sciences (IMLS) in 1974 and the new professional designation was Medical Laboratory Scientific Officer (MLSO). There were seven grades of MLSO: two training grades, Junior A and Junior B, the Basic grade, Senior, Chief, Senior Chief and Principal. Registration with the CPSM was achieved through an oral examination after a suitable period of training. This was a prerequisite for employment at the Basic grade and Fellowship (see below) of the IMLS was usually needed for someone to be considered for promotion to Senior MLSO. In 1988, the grade system was altered to trainee MLSO, then MLSO 1, 2, 3 and 4.

The IMLS assumed its current name of the Institute of Biomedical Science (IBMS) in 1994, allowing the introduction of 'Biomedical Scientist' (BMS) as the professional title. In 2004, the NHS implemented a streamlined structure of pay and conditions called Agenda for Change, which put all staff, apart from doctors and dentists, on the same scales. With this came the current nomenclature of registered practitioner, specialist practitioner, etc., and Bands 5–9 for biomedical scientists in the UK. Over the years, the membership and examination systems have been altered several times in response to the changing needs of the profession. Under the current structure, a person in training could join as an Associate member of the IBMS and progress through Licentiate and then Member, to Fellow during their career. The examination system has been designed to underpin the Agenda for Change grading structure, allowing biomedical scientists to gather evidence of how their professional and scientific knowledge and competency are developing. This is intended to help those who wish to obtain promotion to more senior posts and to take on specialized roles such as quality manager or transfusion practitioner.

1.4 Role of the IBMS as the professional body for biomedical scientists in the 21st Century

The Institute of Biomedical Science (IBMS) is the professional body for biomedical scientists in the UK and its stated aim is 'to promote and develop biomedical science and its practitioners' (www.ibms.org). Most current members joined the Institute while training as biomedical scientists in diagnostic laboratories and continued this association as their careers progressed. Thus, while the membership mainly comprises practitioners in clinical laboratories, staff in specialized diagnostic laboratories, researchers and lecturers are also IBMS members. Academic staff, sales representatives and others who are interested in the work of the Institute are also able to join. The main roles of the IBMS are summarized in Box 1.2.

Box 1.2   Roles of the Institute of Biomedical Science (IBMS)

The main roles of the IBMS are to:

- Set professional standards of conduct and practice for members

- Promote the work of biomedical scientists to the general public

- Represent the issues and concerns of biomedical scientists nationally (for example to the media and at the health departments)

- Advise government and national bodies on all issues which involve biomedical science and affect biomedical scientists

- Set academic educational standards and work with universities to accredit degrees which meet these standards

- Set standards of competency for professional practice and provide means of assessment of these standards, leading to recognized qualifications for members who meet them

- Organize a scheme for continual professional development, which is underpinned by scientific meetings at local and national level

- Publish the monthly *Biomedical Scientist* magazine and the quarterly *British Journal of Biomedical Science*

- Provide grants to support members' research activities

- Offer legal and technical services for members who require them

- Award Chartered Scientist status to suitably qualified senior members

---

The IBMS has a central office in London where full-time members of staff work to assess qualifications of applicants for membership, collect members' subscriptions, organize the Continuing Professional Development (CPD) programme (see Chapter 8), recommend members for Chartered Scientist status, oversee professional examinations, process applications for research grants and coordinate the activities of members at regional and local level (see below). The IBMS publishes a monthly magazine (the *Biomedical Scientist*), which keeps members up to date with news, science and job vacancies, and a quarterly scientific journal (the *British Journal of Biomedical Science*), in which practitioners report research findings. Staff from the central office liaise with government departments (e.g. the Department of Health) and other professional bodies to make sure that biomedical scientists in particular and pathology in general are taken into account in decision making.

Much of the work of promoting and developing the profession is carried out by members, as part of their own professional development. There are regional groups throughout the UK (e.g. North West of England region), plus Ireland, Cyprus, Gibraltar and Hong Kong, and local branches within each region (e.g. Merseyside branch). Members living or working in a particular area can join their local regional group and branch and help to organize scientific and professional meetings and also social events, such as quizzes and outings.

For each specialist area of pathology (e.g. haematology, microbiology) there is a Scientific Advisory Panel (SAP) of practising biomedical scientists, who provide expert advice to the IBMS in their subject area on a variety of issues. Members of these panels are also responsible for organizing the scientific programme in their specialist area at the biennial IBMS Congress, devising continuing professional development activities and setting and assessing IBMS professional qualifications.

There are currently around 16 000 members of the IBMS in the UK and overseas. A biomedical scientist who is a member of the IBMS is showing commitment to their profession in addition to their own personal career development. The Institute depends on ideas and enthusiasm from all its members, to promote the profession and to organize and make presentations at meetings. Table 1.1 gives the categories of IBMS membership as of January 2008 and the requirements to be eligible for each level. There are two types of member: 'corporate' and 'non-corporate'. Those who have the necessary academic and professional qualifications can join the IBMS at the suitable 'corporate' level. Usually a person would start their career as a registered biomedical scientist at Licentiate level and work towards Fellowship over the subsequent 10 or more years. The 'non-corporate' Associate category is available to a range of groups who do not qualify to be Licentiates, Members or Fellows, which includes students. However, students and graduate trainees would be able to apply to become Licentiates after the award of the IBMS Certificate of Competence (Table 1.1). It is clear that biomedical scientists need to gain both academic and professional qualifications to attain higher levels of membership, which are again a mark of the person's dedication to the biomedical science profession.

Table 1.1   IBMS membership categories

IBMS membership category (designation)	Academic qualifications required	Professional qualifications required	Comments
Associate	Enrolment as student on an IBMS-accredited Biomedical Science or Applied Biomedical Science BSc programme	Graduate trainee preparing for IBMS Certificate of Competency	'Non-corporate' membership category
	Or	*Or*	
	Registered practitioner qualified overseas, but not holding necessary qualifications and experience for full IBMS Membership	Associate practitioner, but not eligible for corporate membership	
Licentiate (LIBMS)	BSc Biomedical Science or Applied Biomedical Sciences from IBMS-accredited programme	IBMS Certificate of Competence issued on successful completion of Registration Portfolio	Applicants holding non-IBMS-accredited BSc may meet academic criteria through taking extra modules in an accredited programme
Member (MIBMS)		IBMS Specialist Diploma	Must have been a Licentiate for at least 2 years before applying to become a Member
Fellow (FIBMS)	MSc from IBMS-accredited programme	Higher Specialist Diploma	Must have been a Member for at least 3 years before applying to become a Fellow Prior to 1997, the required qualification was the Fellowship or Special examination
Life member			50 years' service as a member of the IBMS

Members and Fellows who have obtained an MSc, have been in the IBMS for at least 4 years and can demonstrate that they are undertaking continuing professional development (see Chapter 8) are also eligible to apply for Chartered Scientist status. This is indicated by the letter CSci written after a person's name.

1.5 Health Professions Council

The Health Professions Council (HPC) is the regulatory body for 13 health-care professional groups, including Biomedical Scientists, in the UK. The HPC was set up in 2003 under the Health Professions Council Act 2002, to replace the CPSM, which, as mentioned previously, operated under legislation passed in 1960. By this law, a person working in a clinical laboratory, contributing to patients' diagnoses and using the professional title 'biomedical scientist' in the UK, must be registered with the HPC. Its stated aim is to protect the public who use the services of practitioners belonging to the professions it regulates, as listed in Box 1.3 (www.hpc-org.uk). It does this by setting Standards, which outline the competency, education and behaviour expected of registered practitioners (called 'registrants'). These are summarized in Box 1.4. These sets of Standards are common to all registrants, but interpreted by the relevant professional body (e.g. IBMS, Association of Operating Department Practitioners, British Dietetic Association) for the specific tasks and responsibilities each professional group takes in patient care. The HPC register is available in the public domain on its website, so it is possible to check whether any practitioner is registered and to find their registration number. This makes it easy for employers to ascertain that someone to whom they are about to offer a job is a registrant and also allows anyone to lodge a formal complaint against a clearly identified individual.

Box 1.3   Professions currently regulated by the Health Professions Council*

- Arts therapists
- Biomedical scientists
- Chiropodists/Podiatrists
- Clinical scientists
- Dieticians
- Occupational therapists

Box 1.3   (Continued)

- Operating department practitioners

- Orthoptists

- Prosthetists and orthotists

- Paramedics

- Physiotherapists

- Radiographers

- Speech and language therapists

*Other professional groups may be added to this list as new professions can apply to the HPC to be regulated provided that they meet specified criteria.

Box 1.4   Standards published by the Health Professions Council

Standards of Proficiency: Knowledge and competency that a practitioner must demonstrate to be registered; all registrants must maintain this as appropriate to their job role throughout their career.

Standards of Education and Training: Essential elements of an academic programme which must be contained in a university or college course leading to an award which is in whole or part of the requirement to be a registrant.

Standards of Conduct, Performance and Ethics: Code of behaviour and aptitude that registrants are required to meet, in order that colleagues, patients, clients and the general public will be confident in their suitability to practice.

Standards for Continuing Professional Development: Outline of the responsibility that registrants must take to keep themselves up to date in their area of practice.

1.5.1 HPC standards of proficiency

The vocational part of a biomedical scientist's training is designed to produce competent practitioners who meet all the Standards of Proficiency. This is currently achieved through a portfolio of evidence, according to IBMS guidelines (see below). All biomedical scientists who are on the HPC register are required by law to work according to the Standards of Proficiency at all times. Biomedical scientists who were on the CPSM register when the HPC was created were transferred to the HPC register. Therefore, although only recently qualified practitioners will have completed a portfolio to demonstrate their knowledge of and competency in the Standards of Proficiency, more senior staff have to be familiar with the Standards and be responsible for their actions if they do not meet them. About 23 000 people are registered with the HPC to practice as biomedical scientists, though some of them will be working in the private sector in the UK and others are practising in other countries.

In most countries around the world, training courses for laboratory staff combine theoretical knowledge and understanding with practical experience. Although the basic qualification is often a BSc degree, the content can vary, as it depends on the structure of the healthcare service in each country and the responsibilities that scientifically qualified people take on. This means that while practitioners who trained and qualified outside the UK may be admitted on to the HPC register, each case is considered on an individual basis. Similarly, an IBMS-accredited BSc and HPC registration does not automatically allow a UK biomedical scientist to practice in another country. A probationary period working in a clinical laboratory or studying for an extra qualification may be required.

Table 1.2 gives the HPC Standards of Proficiency as they are given to all professional groups. They have then been 'interpreted' by each professional group to make them appropriate to the different types of work that each registrant profession performs. As the professional body, the IBMS worked with the HPC to provide a suitable version for biomedical scientists. These can be found in the *Health Professions Council Standards of Proficiency – Biomedical Scientists* (www.hpc-org.uk). It is worth noting that the standards in Section 1 are concerned with a person's general professional attitude and awareness of codes of conduct, guidelines and the laws which relate to working in a clinical pathology laboratory. Section 2 relates to the application of scientific and technical principles to the specific work that biomedical scientists carry out, and in Section 3, evidence of knowledge and understanding of the theoretical principles behind the main pathology disciplines is asked for, along with a detailed awareness of the importance of health and safety, which underpins all activities in the laboratory.

For pre-registration training purposes, the IBMS has compiled the Registration Portfolio, which can be used both by students on 'coterminous' or 'integrated' degrees and trainees who are full-time employees in a particular department. This provides lists of knowledge and competence necessary to meet each Standard of Proficiency and gives guidelines about collecting evidence. University courses which are IBMS accredited and incorporate the Registration

Table 1.2   Health Professions Council Standards of Proficiency

---

Part 1: Expectations of a health professional

Section 1a: Professional autonomy and accountability

Registrants must:

1a.1	be able to practise within the legal and ethical boundaries of their profession
1a.2	be able to practise in a non-discriminatory manner
1a.3	understand the importance of and be able to maintain confidentiality
1a.4	understand the importance of and be able to obtain informed consent
1a.5	be able to exercise a professional duty of care
1a.6	be able to practise as an autonomous professional exercising their own professional judgement
1a.7	recognize the need for effective self-management of workload and be able to practice accordingly
1a.8	understand the obligation to maintain fitness to practice

Section 1b: Professional relationships

Registrants must:

1b.1	be able to work, where appropriate, in partnership with other professionals, support staff, service users and their relatives and carers
1b.2	be able to contribute effectively to work undertaken as part of a multi-disciplinary team
1b.3	be able to demonstrate effective and appropriate skills in communicating information, advice, instruction and professional opinion to colleagues, service users, their relatives and carers
1b.4	understand the need for effective communication throughout the care of the service user

Part 2: The skills required for application of practice

Section 2a: Identification and assessment of health and social care needs

Registrants must

2a.1	be able to gather appropriate information
2a.2	be able to select and use appropriate assessment techniques
2a.3	be able to undertake or arrange investigations as appropriate
2a.4	be able to analyse and critically evaluate the information collected

Section 2b: Formulation and delivery of plans and strategies for meeting health and social care needs

Registrants must:

2b.1	be able to use research, reasoning and problem solving skills to determine appropriate actions
2b.2	be able to draw on appropriate knowledge and skills in order to make professional judgements
2b.3	be able to formulate specific and appropriate management plans including the setting of timescales
2b.4	be able to conduct appropriate diagnostic or monitoring procedures, treatment, therapy or other actions safely and skilfully
2b.5	be able to maintain records appropriately

Section 2c: Critical evaluation of the impact of, or response to, the registrant's actions

Registrants must:

2c.1	be able to monitor and review the ongoing effectiveness of planned activity and modify it accordingly
2c.2	be able to audit, reflect on and review practice

Part 3: Knowledge, understanding and skills

Registrants must:

3a.1	know and understand the key concepts of the bodies of knowledge which are relevant to their profession-specific practice
3a.2	know how professional principles are expressed and translated into action through a number of different approaches to practice and how to select or modify approaches to meet the needs of an individual, groups or communities
3a.3	understand the need to establish and maintain and safe practice environment

www.hpc-org.uk

Portfolio into their academic programme have flexibility on how they deliver the required training and laboratory experience. Students enrolled on such courses will be advised about how to approach the tasks. For more information on the Registration Portfolio contact the IBMS Registration Department.

Students and trainees who are successful in the IBMS registration portfolio verification are issued with a Certificate of Competence, which is needed to apply to have one's name entered on to the HPC register as a 'registrant practitioner' (although this is subject to satisfactory references and criminal record and health checks). This Certificate is also needed to become a Licentiate member of the IBMS (see Table 1.1), which is important because one cannot be considered for the IBMS Specialist Portfolio unless one is an Institute member.

1.5.2 Role of the HPC in ensuring that practitioners and educators adhere to its standards

Since the HPC standards are published in the public domain, registrants who do not meet them can be called to account, thus protecting the public. A patient, other member of the public or another practitioner can follow a clear complaints procedure if they think that a registrant or academic course is not meeting particular standards. The HPC undertakes to investigate all complaints rigorously. Most concerns are raised about the 'fitness to practice' of a professional registered with the HPC. There is a clear procedure to follow, which makes it fair to both the complainant and the HPC registrant. The issues which can be raised with the HPC about a particular registrant include unprofessional behaviour, incompetence (potentially putting both patients and colleagues at risk), breaking the criminal law (for example, drink driving, possession of class A drugs) or being physically or mentally unfit. The HPC also investigate cases where people have falsely claimed to have obtained the qualifications necessary for entry on to the register, when this is discovered. Similarly, registrants are asked to verify that they are undertaking suitable continuing professional development regularly each time they re-register (see Chapter 8) and potentially fraudulent claims would be looked into. Thus, a registrant can be called to account for not meeting the Standards of Proficiency or the Standards of Conduct, Performance and Ethics in their professional practice. The accusation must be submitted in writing and the person making the complaint must identify themselves to the HPC, so that the matter can be scrutinized thoroughly. The first step is for the complaint to be considered by a panel comprising a number of people, including someone from the same profession as the person under investigation (who can comment on the effect of the registrant's alleged behaviour) and at least one person who is not on the HPC register (who can be more impartial). If this panel decides that there is a case to answer, the matter is passed to either the Conduct and Competence Committee or the Health Committee, as appropriate. At this stage, the allegation is made public – the accused registrant's name and registration number along with a description of the complaint against them are published on the HPC website. A date is also given for the hearing and any person affected by the matter is allowed to attend as an observer.

The outcome of a fitness to practice hearing depends on the situation. Each case is considered individually and the circumstances surrounding the alleged incident or behaviour are taken into account. In some cases, the first panel decides that the HPC registrant has no case to answer. Sometimes, the case is referred to the Conduct and Competence or Health Committee, but they decide not to pursue the matter. When this decision has been made, no further action is taken and the person can resume their full duties at work. If the registrant has done something unwise and unprofessional but there have been no serious consequences (for example, using study and sick leave from one job to work for another employer), they are usually given a 'caution'. This means that they can continue to work as an HPC registrant, but the word 'caution' appears beside their name on the HPC register for a specified length of time. Where the registrant has been struggling to do their job effectively, but has not recognized this (for example, due to illness which has adversely affected their work), they will often have 'conditions of practice' imposed on them. This means that they can continue to work as an HPC registrant, but under supervision and/or after a period of re-training. The HPC would usually review the situation after a suitable length of time (e.g. a year) and decide whether the person is now competent to practice again. In more serious cases, where registrants are suspected of having been working while under the influence of alcohol or drugs or failing to take responsibility for mistakes in their work, they may be suspended from the HPC register for a defined period or even 'struck off' altogether.

1.6 Education and training for biomedical scientists

In the 1970s, clinical pathology departments usually recruited school leavers to train as MLSOs (see Institute of Biomedical Science, above). People joined the laboratory either at 16 years old on leaving compulsory education or, more commonly, at 18 years old, having gained A-levels. Then, as now, few school leavers had a good idea of what the work in diagnostic pathology entailed and they frequently only decided to take up the career after a period of work experience. Trainees were started at the 'Junior A' grade and were sent to college to study part time (on day release) for Ordinary National Certificate and Diploma (ONC/OND) and then Higher National Certificate and Diploma (HNC/HND) in Medical Laboratory Sciences. Within the workplace, they spent several months working in each discipline on rotation, over the course of 2–3 years. Once the period of initial training had been successfully completed, the trainee opted to specialize in one discipline and continued developing their skills and learning more of the theory in that area. They could continue their academic study by taking a BTec or a BSc in their chosen subject, during which time they could be promoted to 'Junior B'. After a further 1–2 years, the trainee would be assessed for registration with the Council for Professions Supplementary to Medicine (see Health Professions Council, above) and became a Basic grade MLSO. During the late 1970s and early 1980s, some laboratories started to recruit graduate trainees, who were started on the Junior B grade and trained exclusively in one discipline. As this became more accepted practice, the opportunity for trainees to rotate through all disciplines became limited and this was eventually stopped.

By the early 1990s, universities began to respond to this preference for graduate recruitment by offering BSc programmes in Biomedical Science, which ran alongside (and in time replaced) the HNC/HND courses. These courses combined teaching of theory with practical training in laboratory skills (Figure 1.2). To ensure that these degree programmes met the profession's requirements, the IBMS set out criteria about the specific knowledge and skills that should be taught in courses, and universities could seek IBMS accreditation for their Honours degree programmes in Biomedical Science. At this time, the pre-registration training for most other healthcare professionals (e.g. nurses, occupational therapists, dieticians) was being changed from a diploma to a degree qualification which included clinical placement experience. In 1995, Biomedical Science followed suit and became a graduate entry profession; this meant that most trainees already had a degree or, if they did not, they were contractually obliged to study for one. The IBMS also laid out minimum standards

Figure 1.2   Students practice clinical microbiology skills in a university practical class

for laboratories who wanted to take on trainees and granted approval as 'training laboratories' only to those who could demonstrate that they met them. Graduate trainees were usually recruited by one of the individual departments within pathology and trained specifically in that discipline. In the early 1990s, the CPSM introduced a 'log book', which was a list of tasks which a registered practitioner in that discipline would be expected to be able to perform. When the trainee could demonstrate an understanding of the theory behind a particular laboratory test (for example, by answering questions about it) and was deemed competent in that task by the colleague who had trained them to perform it, then the relevant section of the log book was signed. This was produced by the candidate at their oral examination as evidence that they had successfully undergone a period of structured training.

Until the mid-2000s, this was the pattern of recruitment and training for biomedical scientists in the UK. Since graduates started as 'trainees', they were out of step with their colleagues in the other graduate professions on the HPC register, who had completed clinical placement training during their degree and were therefore 'registered practitioners' at the time of their graduation. Therefore, the pre-registration training for Biomedical Scientists has now been changed to align it with other professions regulated by the HPC. The key difference is that previously, trainees were expected to acquire detailed, specialized knowledge and competence in a single discipline (for example, so that they could participate in the 'out of hours service') before being assessed for

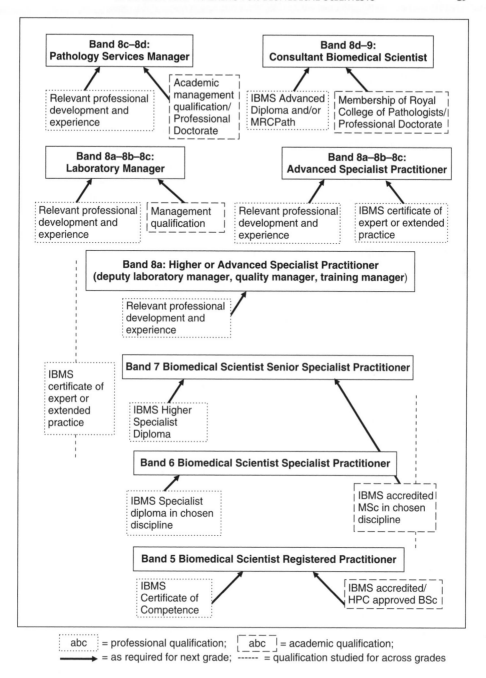

Figure 1.3 Current career pathway for biomedical scientists

registration. Now, it is expected that a newly registrant biomedical scientist will be safe and trustworthy in the laboratory, but will still need training in more depth in the work of the laboratory when they start their first job. This allows for the pre-registration training to give students experience of more than one pathology discipline. Degree programmes for other HPC registrant professions, such as Physiotherapy and Occupational Therapy, include placements in a range of different settings, such as critical care, paediatrics and mental health. For a biomedical scientist, this 'range of different settings' might be the clinical chemistry, medical microbiology and histopathology laboratories. Thus, 'coterminous' or 'integrated' Applied Biomedical Sciences BSc programmes are ideally suited to provide this 'multidisciplinary' training. The accredited degree itself covers all areas of pathology and students usually have the opportunity to observe the work in more than one laboratory while on placement, even if they eventually concentrate on one discipline for the majority of their placement time.

One of the reasons for this change is to make undergraduates more aware of the work that biomedical scientists do within the healthcare service and to introduce them to the career possibilities. Those of us who have found our work as biomedical scientists stimulating and rewarding would also like to see people making deliberate choices to join the biomedical scientist profession, in the same way that they choose to be lawyers, nurses or footballers. Figure 1.3 outlines the current career pathway for biomedical scientists. It shows the qualifications and experience that a newly registered practitioner would need to obtain to become an advanced practitioner, service manager or a consultant biomedical scientist. From this, is it clear that a combination of academic and professional qualifications (plus hard work and dedication!) are required, as illustrated by the career pathway case study described in Box 1.5.

Box 1.5   Career pathway case study

Mary is 34 years old and has worked in the Haematology Department of a teaching hospital in a major city for the last 2 years. She started work as a trainee biomedical scientist at her local hospital, after gaining an IBMS-accredited BSc. Three years ago, Mary passed an IBMS-accredited MSc that she had studied for on a part-time basis at a nearby university. With that qualification and 13 years' experience, she was able to apply for the Band 7 post that she now has at the larger hospital. Mary is working towards taking the IBMS Higher Specialist Diploma (HSD) in haematology, as she would like to take on more senior roles in the future.

She is active in the IBMS, participating in the CPD scheme and organizing monthly lunchtime scientific meetings open to all her colleagues in the Pathology Department and other hospital staff. Therefore, Mary also aspires to obtaining the Fellowship of the IBMS, for which she needs the HSD qualification. Mary has started a distance learning course which is preparing her to take the IBMS Certificate in Extended Practice in Quality Management, since she is interested in eventually working as a pathology quality manager.

1.7 Codes of conduct

Biomedical scientists are bound by law to work safely in the laboratory, process information that they have access to in confidence and treat colleagues with respect (see Chapter 5). They also have two Codes of Conduct with which they must comply. Although it might seem enough to use common sense and accepted practice as guides to behaviour while at work, it is useful for professionals to have codes of conduct for two main reasons: first so that everyone is clear about what is expected of members of that profession, and second so that when someone's behaviour is not satisfactory, it can be pointed out to them unambiguously, with clear suggestions for improvement.

The two Codes of Conduct that biomedical scientists and those training to be biomedical scientists must be aware of and work to at all times, are:

1. the IBMS Code of Professional Conduct (see Box 1.6), as given in Part 1 of *Good Professional Practice for Biomedical Scientists*, 2nd edition (www.ibms.org), which applies to members in all categories including students;

2. the HPC Standards of Conduct, Performance and Ethics, as described in Box 1.4 (www.hpc-org.uk), with which trainees are expected to comply during their training.

Careful study of both codes will reveal that they reinforce the legal requirements to which biomedical scientists must adhere, for example, confidentiality and health and safety legislation (see Chapter 5). However, they are also both concerned with general attitudes, expecting that biomedical scientists will do their best at all times whatever the task in hand. They encourage practitioners to behave politely and respectfully to everyone they meet in a professional capacity. IBMS members and HPC registrants are exhorted not to bring themselves, their profession or the professional body into disrepute. It is important to know these codes because learning to take one's professional duties seriously is a key part of training to be a biomedical scientist.

Box 1.6   The IBMS Code of Professional Conduct

All members of the Institute of Biomedical Science shall always:

1. Exercise their professional judgement, skill and care to the best of their ability.

2. Fulfil their professional role with integrity, refraining from its misuse to the detriment of patients, employers or professional colleagues.

3. Seek to safeguard patients and others, particularly in relation to health and safety.

4. Treat with discretion all confidential and other information requiring protection and avoid disclosure to any unauthorized person the result of any investigation or other information of a personal or confidential nature gained in the practice of their profession.

5. Act in good faith towards those with whom they stand in professional relationship and conduct themselves so as to uphold the reputation of their profession.

6. Strive to maintain, improve and update their professional knowledge and skill.

7. Promote the study and development of biomedical science and the education and training of biomedical scientists.

www.ibms.org

1.8 Conclusion

This chapter has shown that biomedical scientists have a long tradition of working to support patient care, through diagnostic work, research and teaching. They have a legal requirement to be registered to practice in the UK and also a duty to work to the highest standards of professional conduct. Biomedical scientists have opportunities to enhance their skills and knowledge, through training in new areas and taking recognized qualifications, thus leading to a challenging and rewarding career.

Quick quiz

1. Name the three main ways in which pathology contributes to patient care.

2. List the specialist disciplines within pathology.

3. Distinguish between the functions of the IBMS and the HPC.

4. What qualifications are required to become a Fellow of the IBMS?

5. Name the four HPC sets of 'Standards'.

6. State four roles of the IBMS.

7. Why is it necessary for professionals to have codes of conduct?

8. Outline the procedure which is followed if a member of the public makes a formal complaint against an HPC registered practitioner.

Coursework exercises

1. Visit the IBMS website (www.ibms.org) and investigate the full range of professional qualifications that are available to members. Find out in which disciplines one can take the Higher Specialist Diploma and which Certificates of Extended Practice are currently available.

2. Visit the HPC website (www.hpc-org.uk) and find out the registration number of a colleague (with their permission). Find, download and carefully read the Standards of Conduct, Competency and Performance. Think about your experiences in your training laboratory(ies) so far and write a short piece of reflection on whether you have been keeping these standards and how your conduct could improve.

3. Set up interviews with five biomedical scientists – of various ages and grades – and ask them about their training and how they achieved their academic and professional qualifications. Think carefully about your questions so that you obtain useful information from your interviewees. Record your findings as a series of bullet points or in a table. Then write a short reflection on how your attitude to your own career as a biomedical scientist has changed after interviewing these people.

Suggested references

Health Professions Council website: www.hpc-org.uk.
Health Professions Council (2003). *Standards of Proficiency – Biomedical Scientists*, London: HPC, or later edition.

Health Professions Council (2004). *Standards of Conduct Performance and Ethics*, London: HPC, or later edition.

Institute of Biomedical Science website: www.ibms.org.

Institute of Biomedical Science (2005). *Good Professional Practice for Biomedical Scientists*, 2nd edn. London: IBMS, or later edition.

2

Organization of pathology departments and the role of pathology in healthcare

2.1 Introduction

Most modern pathology departments in the UK are large and complex, employing 100 or more individual members of staff, performing a wide variety of tasks. They contain expensive and sometimes extensive pieces of equipment, often situated across a number of geographically separated sites. There are a number of different staff groups, who are trained and work in specific disciplines. Clearly, for this arrangement to be successful, it needs careful organization, strong leadership and effective management. A well-run pathology department can be an asset to the local healthcare service and biomedical scientists can play a vital part in this. This chapter describes the overall organization of pathology departments, the roles of the different staff groups that work together in a typical pathology laboratory and the working relationship between pathology and the other key services of the healthcare system. The ways in which pathology is continually improving its function and service are also discussed, with consideration in particular of the importance of evidence-based laboratory medicine and the application of point of care testing.

2.2 Organization of pathology departments

Hospital pathology departments have grown from occupying one small room at the turn of the 20th century to now taking up whole sections of buildings, over several floors and often on more than one site. Developments in science and technology led to laboratory staff beginning to specialize in one area and thus the separate pathology disciplines (e.g. haematology and clinical chemistry) emerged; staff and equipment concerned with a particular discipline tended to work in defined areas. Eventually, this meant that discrete suites of rooms were

An Introduction to Biomedical Science in Professional and Clinical Practice   Sarah J. Pitt and James M. Cunningham
© 2009 John Wiley & Sons, Ltd

allocated to each discipline. Initial training was still multidisciplinary, until the 1980s, by which time trainees were usually recruited to one discipline. Now in the 21st century, with the advent of large-scale automation and the use of molecular biology in routine diagnostics, that allow the same technology to be applied to tests across all the pathology disciplines, some of these barriers are being broken down. The IBMS-accredited Biomedical Science BSc programmes are designed to introduce students to all the specialist areas in pathology and to encourage them to understand how the disciplines can work together to the benefit of the patient.

A typical pathology service comprises the main disciplines of clinical chemistry, haematology and transfusion science, histopathology and cytopathology and microbiology. In some larger pathology departments – for example, those associated with teaching hospitals or specialist centres — there may be separate laboratories concerned with immunology, genetics, mycology, parasitology or virology. These are discussed in Chapter 6. The work of all these laboratories is supported by the specimen reception and information technology sections and overseen by service managers. Some of the laboratory aspects of public health medicine are assumed by the Health Protection Agency; the Blood Transfusion Service also has its own laboratories. The departments within these organizations are arranged in a similar way to hospital laboratories and have a similar mix of staff groups to that found in a general pathology department; these are described in detail in the following section.

2.3 Staff groups within pathology

A wide range of staff groups, with varying remits and expertise, are required for the successful running of a modern pathology department (see Box 2.1). All jobs are essential and biomedical scientists should value and support the work done by all of their colleagues. For example, although the staff working in specimen reception do not need specialist scientific knowledge and qualifications, their job is central to pathology. If a specimen is not clearly identified as having been taken from the correct patient or it has been collected or stored incorrectly, then there is no point in processing it. The result would not be helpful for the patient and indeed may be dangerous! It is important to bear in mind that no matter how sophisticated the technology or sensitive the test is, the result is meaningless if it does not apply to the right person. Therefore, specimen reception staff should take time and care over their work and it should be appreciated.

Box 2.1 lists all the staff groups involved in contributing to the running of a high-quality pathology service. Although not all departments have representatives of all the groups shown among their staff, it is worthwhile considering the role that people in each of these jobs plays towards providing accurate, clinically useful results in time to help the patient.

Box 2.1   Staff groups in pathology

- Cleaners

- Porters

- Associate practitioners

  – Phlebotomists, medical laboratory assistants, medical technical officers, cytology screeners

- Healthcare scientists

  – Biomedical scientists, clinical scientists

- Medically qualified staff

- Specialist nurses

- Clerical and administrative assistants

- IT support staff

2.3.1 Cleaners

The cleaning staff are not usually employed directly as part of the pathology staff, but their job is nevertheless important within the department. They clean and polish the floors in the laboratories and offices and also dust some of the work surfaces in areas where patients' samples have not been handled. They empty non-hazardous waste from the appropriate bins and are responsible for ensuring that there is always enough soap and hand towels for staff to wash their hands in the laboratory and rest areas. This job is important because the laboratories and offices need to be as clean and tidy as possible to allow people to work effectively. It is also much nicer to have a pleasant working environment.

2.3.2 Porters and drivers

The staff responsible for transporting items to and from pathology are also employed and managed from outside pathology. The porters' job is to collect patients' specimens from wards and outpatient departments within the hospital and then take them to the pathology department. Vans are used to transport specimens which have been taken at general practitioners' surgeries back to the hospital laboratory; similarly, when the pathology department is located on several sites and certain tests are only done at one location (see below), drivers are employed to take specimens between sites. Porters and drivers also take unused containers, such as blood collection tubes, swabs, microbiological culture tubes and glass slides, back to the wards and clinics and general practice surgeries. Although results are made available electronically, paper copies are still printed and porters sometimes take them to the wards when required. In many hospitals, there is a 'pod' system, which is used for transporting specimens, fresh containers and results between pathology and other departments. This consists of a series of pipes, into which a sealed 'pod' containing specimens or containers is placed and moved between departments under pressure within a vacuum. This can be very efficient, but such a system is only useful when the laboratory is in the same building as the wards. It is important that the transporting job is done efficiently, so that specimens arrive in the laboratory in good time to be processed and clean containers are always available for the staff who examine and take samples from patients.

2.3.3 Phlebotomists

Due to the sheer volume of work involved in routine blood collection, specially trained staff are employed to do this in most hospitals. Phlebotomists visit the wards each day and collect blood from patients who need it as part of their scheduled ongoing care, according to the instructions of the clinical staff looking after them. They also play a similar role in primary care, by holding regular sessions at general practice surgeries for patients whose blood tests are not urgent. The samples are then taken or sent back to the pathology department for the required tests. As they take so many blood samples, phlebotomists tend to be very skilled at this task and at reassuring patients. They must also be very meticulous about using the correct blood collection tubes, due to the range of additives which are used depending on which test is to be carried out on the sample. This role is valuable because when the correct specimen is collected from the right patient competently, it helps to ensure that staff on the wards receive the results in good time to treat their patients appropriately.

2.3.4 Laboratory Assistants and Associate Practitioners

Assistant and associate practitioners take on a variety of roles which help biomedical scientists in their work. They do one of the most important jobs in the whole system, which is the processing of samples as they come into pathology. This involves receiving specimens which are delivered to the pathology department, either by pod, porter or the patients themselves. The assistant practitioners then categorize specimens which arrive according to whether they are urgent or non-urgent and decide the section of the laboratory to which they should be sent. They check that sample bottles and pots have not been broken or leaked during their passage to the laboratory. They must also verify that the details given on the request form match those on the specimen container (see Chapter 5). It is usually the job of specimen reception staff to record the relevant details on the computer, which allows the sample to be allocated a laboratory number and thus be processed. Working under the supervision of biomedical scientists, assistant practitioners in laboratories carry out a variety of tasks including separating, labelling and storing serum and tissue samples, loading samples on to analysers, making up stocks of chemicals, quality control samples and microbiological culture media and monitoring equipment such as refrigerators and incubators. All these jobs are extremely important to the smooth running of the laboratory. When they are done well, it reduces the 'turnaround time' (see Chapter 5) for the samples and allows registered colleagues to concentrate on more complex parts of the analyses. Another example of assistant practitioners are mortuary technicians, who run the mortuary facilities and assist anatomical pathologists when doing post mortem examinations.

It is not necessary to have any specific academic qualifications to start work as an assistant practitioner. However, a BTEC National Vocational Qualification (NVQ) in Clinical Laboratory Support has been introduced to allow staff in these posts to learn underlying theory and gain credit for their practical experience. Foundation degrees in biomedical sciences are also being developed, to support the training of staff wishing to further their careers by taking on increased responsibility. This qualification is intended to support staff who wish to take on the role of an associate practitioner. In future, staff at this grade are likely to perform some routine tasks which are currently performed only by biomedical scientists.

2.3.5 Cytoscreeners

Cytoscreeners spend their working day examining slides from cervical smears, to look for signs of abnormalities. They are comprehensively trained for this specific task, undertaking a 2 year course leading to assessment for the National Health Service Cervical Screening Programme Certificate of Cervical Cytology. Their skills and competency are regularly reassessed to make sure that they are maintaining the required standard in their work. Because of this, they can do the job very efficiently and refer any slides where the cells do not look normal on to biomedical scientists or medical staff. Through accurate detection of possible cancerous and pre-cancerous cells in smears, cytoscreeners help to identify patients who need medical intervention early, thus saving lives.

2.3.6 Healthcare scientists

The group of 'healthcare scientists' comprises biomedical scientists and clinical scientists. Although these two professions currently have discrete training and career structures, their roles often overlap. At the time of writing, changes to the training and job descriptions for healthcare scientists are being proposed, which would remove that distinction; the aim is eventually to have one professional group. The roles are described separately here, to reflect the present situation:

2.3.6.1 Biomedical scientists

Biomedical scientists undertake the bulk of the work involved in processing routine specimens. This involves selecting the appropriate markers to test for and the techniques to use on a particular sample, carrying out the tests and interpreting the scientific implications of the results. The nature of the work depends on the discipline they are working in (see Chapter 6). Suitably qualified and experienced biomedical scientists take on line management roles such as section leader, training officer, quality officer or health and safety officer. Biomedical scientists also participate in research projects, to help improve the service delivered to patients. Examples of research include evaluating new testing methods and trying new systems of working. Sometimes projects are undertaken as part of the requirement for a higher academic qualification such as an MSc.

As registered practitioners, biomedical scientists can work autonomously and this sometimes means alone – for example, providing 'out of hours service' or working in primary care. They take responsibility for their work as professionals and also that of trainees and assistant practitioners working with them. This means that they should work to the best of their ability at all times, must train and supervise junior staff properly and keep up to date through continual professional development (see Chapter 8).

Many biomedical scientists find that opportunities arise for them to complement their laboratory benchwork by developing their skills in specific areas. For example, in addition to teaching junior staff about laboratory techniques and scientific theory, some biomedical scientists lecture at universities on Biomedical Sciences BSc and MSc courses or act as tutors for distance learning programmes. They often undertake extra qualifications in teaching, such as a postgraduate certificate or the IBMS Certificate of Extended Practice in Training, to help with this role. Other biomedical scientists find that they have a particular interest in quality management and want to contribute towards ensuring that their department works to the highest standards of technical expertise and service. The IBMS Certificate of Extended Practice in Quality Management is available to help with their professional development in this area. Another option is to specialize in a clinical area such as infection control, which is again supported by the IBMS Certificate of Biomedical Practice in Infection Control. The working days for a recently registered Biomedical Scientist and a Laboratory Manager described in Boxes 2.2 and 2.3 illustrate the wide range of roles and responsibilities that can be experienced in a typical pathology laboratory setting.

Box 2.2   A typical day for a registered Biomedical Scientist

Peter is a registered biomedical scientist and he works in the Haematology Department within the pathology service of a large district general hospital. His typical working day begins at 8.45 a.m. At the moment, he is working in the routine section and his main duties are running full blood counts (FBCs) on the analyser and putting up erythrocyte sedimentation rate (ESR) tests. His first task is to check that the machine is functioning properly, check that the levels of all reagents within the analyser are acceptable and to run the quality control samples through the analyser. When everything is ready, Peter goes to specimen reception to collect his first batch of samples.

There is usually a steady stream of samples for the FBC and ESR tests during the day. They usually comprise a mixture of routine requests and specimens taken from patients in hospital departments such as A&E and the Intensive Care Unit, which require urgent attention. In the morning, most of the routine work comes from hospital wards; later in the day, these samples are joined by those arriving from general practitioner surgeries.

Peter confirms that, for each sample, all the required details are on the request form and that the information on the blood bottle tallies with the form. He also checks that a sample tube containing a suitable anticoagulant has been used and that it has been filled to the correct level, so that the blood has not clotted. He then verifies that all the samples he is intending to process have been booked in by one of his colleagues in the specimen reception and have been assigned a laboratory number.

Peter loads the samples on the analyser, using the laboratory number to identify the position of each specimen in the machine. When he is sure that everything is running smoothly, he takes a 15 minute tea break. Before he leaves the laboratory, he tells his section leader that he is going for tea and also mentions how far he has progressed with the morning's work. Peter takes his break with his friends from other parts of the pathology department, in the main pathology tea room.

Box 2.2   (Continued)

When the results are ready, Peter enters them into the reporting system. If there are any unusual results (for example, high or low white blood cell count), he discusses them with the specialist or senior specialist biomedical scientist in his section. Sometimes, a thin blood film must be prepared which can be examined microscopically; this might be necessary to determine the morphology of the red cells or the type of white blood cell which is present in unusually high amounts.

Peter continues to work steadily in this pattern during the day. Today, he attends a staff meeting chaired by Jane, his laboratory manager, at 11 a.m. and takes his lunch break at 12.00, as allocated by his section leader. He is always careful not to take longer than the agreed amount of time for his break, as he is conscious that if he is late back, that means that someone else's break is delayed. His lateness might also mean that is takes longer than it should for results to be issued, which could affect patient care. Peter takes his afternoon tea break at 3.15 p.m. and, on the way back to the laboratory, he goes to his laboratory manager's office to request some leave.

If Peter encounters any problems, he mentions them to a more senior colleague as soon as possible. A very unusual test result might need to be brought to the attention of the consultant haematologist. If the analyser is not working properly, it may be necessary to call the manufacturer, to ask for an engineer to visit.

The routine working day finishes at 5.30 p.m. Peter never goes home without checking that all the work he was responsible for has been done and that his section leader is happy for him to go. This level of communication is crucial for effective team work with in the department and helps ensure that a good service is provided for the patients.

This evening, Peter is keen to leave on time because he has to be in a lecture theatre at another hospital by 6 p.m. to attend a lecture, organized by the local branch of the IBMS and given by an invited speaker, about haemoglobinopathies. The talk starts at 7 p.m., but as the meeting has been sponsored by one of the private companies which supplies equipment and reagents to all the haematology laboratories in the area, the organizers have been able to provide food and drink for those attending. Peter enjoys these events as he gets a chance to catch up with all the news from his friends and colleagues working in other hospitals, as well as learning more about biomedical science.

Box 2.3 A typical day for a laboratory manager

Jane is the laboratory manager of the Haematology Department where Peter works as a registered biomedical scientist. Her working day usually begins at 8 a.m., when there is a chance to catch up on e-mails, phone messages and any relevant paperwork before meetings start.

This morning at 9.30 a.m., Jane and the Consultant Haematologist meet to talk through some technical and management problems which have arisen in the Haematology Department, as they do regularly. Although she often speaks to the Pathology Services Manager during a normal working day, they also have formal meetings. Tomorrow Jane and the other laboratory managers are due to have their weekly meeting with the Pathology Services Manager; they will consider how to approach any problems which could affect the running of the whole of pathology and discuss the pathology budget.

At 11 a.m., there is the monthly staff meeting, during which all members of the laboratory team, including Peter, are encouraged to ask questions or suggest solutions to problems. Today the results of the latest External Quality Assessment exercises are discussed. The Haematology Department has performed well in these assessments and Jane thanks the staff for their hard work. One of the senior specialist biomedical scientists presents the results of an audit which she has recently conducted on blood samples. She did this work because she had noticed that a lot of blood samples arriving in the laboratory could not be processed because they had been collected incorrectly. The audit showed that they were mostly coming from a single general practitioners' surgery, which Jane has agreed to contact later today. Although it is still several months away, Jane asks all staff to think about possible venues for the Christmas party.

Between 12 and 3 p.m., Jane attends a meeting of the local Pathology Network, which is held in a board room in the main hospital. Pathology Services Managers and Laboratory Managers from several neighbouring Trusts have come and they talk about how they can cooperate to improve the pathology service in the region. The main topic under discussion is how to set up a regional molecular diagnostics laboratory. The managers consider what expertise they have in this area among their existing staff and whether they might need to create a new post for a person to oversee the running of the laboratory and training of staff. They also talk about possible sites for the laboratory and how easy it would be to transport specimens around the region.

Box 2.3 (Continued)

Jane is back in her office in the Haematology Department by 3.10 p.m.; at 3.15 p.m., Peter comes to request some leave, which she is authorizes. One of the senior specialist biomedical scientists then comes to inform her that an order of reagents for the full blood count (FBC) analyser has not arrived today as expected. Jane telephones the company which supplies these reagents to find out what has happened. It transpires that a major fault with one of the delivery lorries has meant that there are some delays, but the person she is communicating with assures her that the order will arrive later that afternoon.

At 4 p.m., Jane and the Consultant Haematologist go to the Oncology Department for a meeting with the senior medical and nursing staff to discuss the quality of the service which the haematology department has been providing.

There are no major problems and when Jane returns to her office, she is informed by a member of the laboratory staff that the reagents for the FBC analyser have arrived. She contacts the manufacturer to confirm the arrival of the order and then leaves the department. She does not go straight home, however, because the local branch of the IBMS have organized a lecture on haemoglobinopathies which is taking place at another hospital and she decides to attend this event. She is pleased to note that Peter and several other members of her staff have also chosen to come to the lecture.

2.3.6.2 Clinical scientists

Some departments employ clinical scientists to take on specialized scientific roles, particularly research and development. Clinical scientists often undertake higher research degrees (MPhil, PhD) as part of their training and develop expertise in a specific area of research. They are also involved with the teaching and training of colleagues within the laboratory. Some senior clinical scientists train for and sit the examinations for Membership of the Royal College of Pathologists, which gives them the appropriate training and qualifications to give some clinical interpretation and advice. At present there are a limited number of posts for clinical scientists and they do have a distinct role and career structure from biomedical scientists, although they are also HPC registered practitioners. However, within the NHS Healthcare Scientist career framework, the boundaries between the two roles at more senior levels are becoming blurred. For example, biomedical scientists who are advanced practitioners provide clinical interpretation and those studying for a professional doctorate will be doing high-level research.

2.3.7 Medical staff

Medical registrars specialize in a particular pathology discipline during their postgraduate training and thus each department has at least one medical consultant. The preparation towards a consultant position includes the examinations for the Membership of the Royal College of Pathologists and a professional doctorate, the MD. Some medical doctors have also completed a BSc in a pathology discipline (e.g. microbiology, biochemistry or biomedical sciences) during their undergraduate studies during an 'intercalating year'. The medical practitioners' role within a pathology department, of any discipline, is to give clinical interpretation of laboratory results for the doctors on the wards and in general practice. They also examine patients and liaise with medical and other colleagues, to ensure that patients are treated and managed appropriately. Consultants also take ultimate responsibility for the results which are sent out by their laboratory, which means that they should have a good working relationship with the laboratory manager and trust him/her to recruit, train and supervise staff correctly. Because they have experience of dealing with patients directly, they can help to inform decisions about which markers to test for and which test requests should be considered 'urgent'. Their insight can also be useful in interpreting results, particularly in unusual or difficult cases.

2.3.8 Nursing staff

Some nurses specialize in areas related to pathology, such as blood transfusion and infection control, and therefore liaise closely with the laboratory staff. In line with most healthcare professional groups, the pre-registration training for nurses is now at degree level. Specialist nurses will have developed an interest in a certain area and gained relevant experience and qualifications as part of their continuing professional development. These nurses fulfil a valuable role by working with their colleagues on the wards, to implement correct procedures which minimize the risk of a patient being given the wrong type of blood during transfusion or the spread of infection between patients. When there are problems, the specialist nurse can help to investigate the causes effectively because they are familiar with the ward environment. They can also work with the pathology laboratory staff to draw up new guidelines to prevent future incidents where necessary.

2.3.9 Clerical and administrative staff

All departments rely on their clerical and administrative staff to ensure that the personnel and documentation are correctly and efficiently organized. In some laboratories, although sorting out the specimens as they arrive is done by the assistant practitioners and biomedical scientists, entering patient details on to the computer system is done by clerical staff. They may also have the job of printing off and sending out paper copies of test results to wards and general practice surgeries. Users of the pathology service can usually access test results electronically, particularly when they are working within the same NHS Trust as the laboratory. However, for urgent results or when the computer is not available, users often phone the laboratory. It is appropriate for biomedical scientists or medical staff to contact colleagues with urgent or unusual results. However, many departments consider that laboratory staff should not be taken away from their main tasks to confirm routine results which have been validated and posted on the computer system; hence clerical staff can take on this role.

Administrative office staff also type up documents such as letters and reports (for example, reports from pathologists' records of histological examination of tissues), file documents, keep records up to date (for example, staff attendance, sickness and leave records) and generally support the manager in the organization of the department, as necessary. Hence the jobs of clerical and administrative staff are central to the smooth running of a pathology department.

2.3.10 Information technology staff

Every part of a modern pathology service, from communicating information, to recording details of specimens before testing, running analysers and reporting results, is dependent on computer systems; this means that each department needs information technology (IT) support from specialized members of staff. It is important that the IT staff have an awareness of how the pathology service works, in addition to a particular interest in computers. Hence these roles are often filled by people who have trained and practised as biomedical scientists. Some of the computer systems used in the laboratory are to support analysers and collate their results; these commonly link into the main systems where results can be authorized and reported so that they are available for users to access. This means that if either system breaks down, the work of the department is severely compromised. The expertise of the IT support staff is vital to minimize the occurrence of problems and to troubleshoot when they do happen. Similarly, computer hardware and software continually need upgrading and the IT staff implement the necessary changes and train other staff to use the new versions.

2.3.11 Laboratory and service managers

It is the laboratory manager's job to ensure that all staff are appropriately qualified, properly trained, happy and healthy enough to do their various jobs well. They do this effectively by supporting and encouraging each individual staff member. They also delegate some tasks and decisions to line managers, who must be therefore be aware of the Trust and departmental policies on management issues. Similarly, the pathology services manager oversees the work of each laboratory manager, usually by providing guidance and advice where necessary. Laboratory and service managers have responsibilities for the large budgets required to buy equipment and reagents and also to cover staff salary and training costs. The pathology services manager's role is sometimes seen as more suited to someone with a business background. Hence although the person in this job may be a biomedical scientist with further training in management and business practices, they also may have a background in another healthcare profession or may have come from outside the health service altogether. Biomedical and clinical scientists in management roles and medical directors liaise with staff from other hospital departments about the running of the hospital. They also consult with users of the pathology service, to ensure that the tests offered and the 'turnaround times' for results are clinically relevant. Although managers inevitably spend little or no time working at the bench, they still have an understanding of the jobs done by all their staff. Their days may be appear to taken up by paperwork and meetings, but a good manager allows their staff to do their jobs to the best of their abilities by ensuring that everyone feels part of the team and has the equipment, resources and training that they need.

Figure 2.1 shows the typical configuration of staff in a pathology department. It shows the different grades and hierarchy for each staff group and the lines of accountability. For example, a senior specialist biomedical scientist is responsible for managing specialist biomedical scientists, trainees and Assistant Practitioners in their section. They are accountable for their own actions and the running of their section to the laboratory manager (or deputy laboratory manager, as appropriate to the individual department).

2.4 Role of pathology in healthcare

Since about 70% of clinical diagnoses rely on pathology test results, staff in pathology have a great impact on patient care. The involvement of biomedical scientists in patient care is often unrecognized by patients and inadequately understood by other healthcare professionals. However, the contribution that can be made by practitioners who are using their training and expertise as scientists, to think objectively and consider the evidence systematically, is invaluable (see Section 2.6).

Laboratory testing needs to be cost effective in terms of both time and money and it is only worth doing the test at all if it will be useful in patient care. Hence the most accurate test must be performed, to as high a standard as possible, on

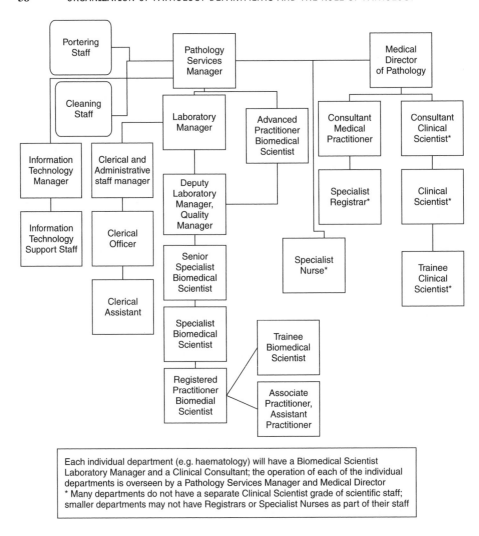

Figure 2.1   Staff groups that may be found in a Pathology Department and lines of accountability

the correct samples, from the right patient and the results must be made available in time to be clinically useful. The laboratory service must therefore be organized in such as way as to achieve this. Most laboratories have 'core' working hours when the highest number of staff are in the department and the bulk of the routine work is carried out. Outside these hours, staff in some disciplines cover shifts (such as evening or night duty) or have a person 'on call' who is required to come to the laboratory at any time to undertake tests where the results are needed to treat a patient with a medical emergency (for example, examining the cerebrospinal fluid from a child with meningitis).

The laboratory manager and medical consultant will liaise with their colleagues on the wards and in general practitioners' surgeries to agree which tests should be done urgently, on demand, and which results are more routine. They will agree 'turnaround times' for each test, which is the maximum length of time between the laboratory receiving the specimen and the result being made available to the staff treating the patient. In some cases, this might be a matter of hours (for example, tests for biochemical markers on patients admitted to Accident and Emergency with acute chest pain), whereas in others it might be weeks (for example, detection of anti-Rubella antibodies in antenatal patients). In the former example, the patient is acutely ill and an urgent assessment of their condition is needed to ascertain whether they require immediate, potentially life-saving, treatment. In the latter, the blood sample for the test is usually taken at the first routine antenatal appointment and discussed with the patient at the next one, some weeks later. In the now rare situation of a pregnant patient, who does not know her Rubella status, coming into contact with someone with acute German measles, the case would become a priority and the relevant test results would be provided urgently.

Since the late 1990s, it has become increasingly common in the UK for groups of two or more individual hospital NHS Trusts in a particular geographical area to merge, forming a larger Trust, under the management of a single Chief Executive. The idea is to concentrate resources and expertise and to encourage colleagues to work together to improve the overall standard of care for all patients in that district. One of the consequences of this has been the 'centralization' of the laboratory service. In some cases all the routine diagnostic testing work is done on one site within the Trust, with a small 'hot' laboratory, providing rapid results for urgent cases only, in each of the other hospitals. Another common situation is for the pathology department to be divided across sites, so that the laboratory space at one hospital within the Trust is adapted to perform all the work that is received for testing in one or two specific disciplines, while the rest is done in another set of laboratories at a different site. This is considered to be more efficient, since, for example, the money used to run two separate sets of analysers can be pooled to obtain one larger and newer machine. There is one overall Pathology Services Manager, which is intended to make the support of staff and the overall decision-making processes more effective. In addition to this closer working at Trust level, a recent innovation is Managed Networks (see Section 2.7), in which Trusts cooperate to provide specialized pathology services across a given area.

Clearly, these systems entail a high level of organization, and effective communication between members of staff is essential for them to work well. The provision of a good quality laboratory service is only possible when each person in the department is competent to do their allotted tasks, feels confident about the level of responsibility they have and is valued as a member of the team. Regardless of the qualifications necessary or the salary each commands,

every job is as important as the others, as described above. Careful consideration of the case of centralized pathology services will show that the drivers and
porters who collect and deliver specimens to and from the various sites are as
essential as the consultants who give clinical interpretations of unusual results.

2.5 Users of the service

Clinical staff, who use pathology test results to guide management of their
patients, are mostly concerned that the right result arrives for the correct patient
in good time to affect the diagnosis and treatment. Managers in pathology usually work very closely with colleagues in the wards and in primary care to ensure
that appropriate laboratory tests are offered, to cover all relevant clinical areas,
in a cost-effective way. This may mean that in a general hospital microbiology laboratory, where a test for anti-rabies antibodies is requested a few times
per year, it does not make economic sense to purchase the necessary reagents
and train the staff to perform the assay. The laboratory can 'buy in' the testing
service, by sending the blood samples to the reference laboratory for analysis. However, where a particular hospital offers specialist services, for example,
for patients with HIV/AIDS, the pathology provision has usually grown with
demand. In such a situation, there would typically be full virology and immunology provision, including monitoring of viral load and $CD4^+$ counts, readily
available within the same NHS Trust. This would have expanded as the range
and accuracy of laboratory tests for diagnosis and monitoring of HIV/AIDS
have increased, over the last 20 years.

Box 2.4 lists the main users of a typical diagnostic pathology service. Staff in
these departments rely on pathology test results being provided accurately and
on time, in order to treat their patients effectively. Some examples are discussed
in more detail below.

Box 2.4   Examples of users of the pathology service

- Accident and Emergency

- Intensive care

- Medical wards, e.g.

  - Endocrinology

  - Gastroenterology

- Cancer services and Oncology wards

- Maternity services

- Paediatrics
- Surgical wards, e.g.
  - General surgery
  - Gynaecology
  - Ear nose and throat
- Outpatient services for patients with conditions which they manage at home, e.g.
  - Warfarin clinic
  - Diabetic clinic
- Genitourinary medicine department
- Primary care, e.g.
  - General practitioners' surgeries
  - Community midwifery services
- Occupational health

2.5.1 Accident and emergency (A&E)

It is very difficult to predict when patients will arrive at the A&E Department and what symptoms they will present with when they do. However, it is clear that a diagnosis must be made as soon as possible, in order to prioritize patients according to the seriousness of their condition. At the time of writing, A&E Departments are also constrained by the '4 hour target', which means that the Department of Health requires that everyone who arrives in the department must have been seen and either discharged from the hospital or moved to a bed within 4 hours. Pathology has a vital role to play in the process. For example, when a person has been experiencing severe chest pain, nausea and dizziness, bio-chemical test results for creatinine kinase and troponin levels can determine whether the symptoms were caused by an acute myocardial infarction and, if so, how severe the heart attack was. Similarly, it would be very important to determine whether a teenager with a headache, pyrexia, stiff neck and pho-tophobia had bacterial meningitis, so that the appropriate antibiotic treatment could be started at once. A motorcyclist who has been involved in a road traffic accident may require emergency surgery and postoperative blood transfusion, which would require the blood grouping and cross-matching procedures to be done urgently. In all these cases, biomedical scientists will have been closely involved in the laboratory diagnosis; if the tests have been performed to the highest standards and in good time, then this could contribute to helping the patient.

2.5.2 General hospital wards

As part of any preoperative assessment, a patient will have blood tests, including a full blood count and routine urea and electrolyte tests, to make sure they are well enough to cope with the effects of the procedure itself and the anaesthetic. Depending on the nature of the operation, they might also need a coagulation test, to ensure that their clotting time is adequate. When the patient is seriously ill and the operation is likely to be extensive and take several hours, there is a danger of blood loss; in this case, their blood group will be determined and preparations made to perform cross-matching with donor blood should it become necessary (the so-called 'group and save' procedure). Clearly, the operation cannot start without the information that the results from pathology provide. It is sometimes necessary to make a diagnosis through histological examination of a sample of a patient's tissue during an operation (for example, to diagnose malignancy in a tumour). In this case, a frozen section is prepared by the biomedical scientists in the histology department (see Chapter 6) and it is examined as a matter of priority by a consultant histopathologist, in order to help the surgeons decide how to proceed with the operation. Postoperatively, patients' progress is monitored biochemically and haematologically. The microbiology department contributes to care on a surgical ward by diagnosing the causative organisms of any infections (such as respiratory infections) and recommending the appropriate antibiotic treatment. Biomedical scientists also carry out tests to monitor the concentration of some antibiotics in patients' blood, when high levels can be dangerous (such as gentamycin).

Patients with general medical conditions or who are in hospital for treatments not involving surgery, such as chemotherapy, also benefit from the work of biomedical scientists. For example, if a patient was admitted to hospital for investigations of acute jaundice, a range of laboratory tests would be done to help make the diagnosis. Liver function tests, performed in biochemistry, would help to decide the extent of the liver damage and possible causes of it, such as ingestion of a toxin, overdose of a drug or alcohol or an infectious agent. Sometimes, the results are needed quickly, but not urgently; nevertheless, the work of the staff in pathology is important to help make the diagnosis, monitor the patient's progress during treatment and minimize the length of their stay in hospital.

2.5.3 Genitourinary medicine clinic

Patients who attend the genitourinary medicine (GUM) clinic are often very worried and embarrassed about the possibility that they might have acquired a sexually transmitted disease. The clinic staff therefore aim to advise patients and if necessary begin treatment for any infections on the day of the appointment. Blood is sent to the microbiology or virology laboratory to test for antibodies and (where appropriate) antigens for organisms such as Human Immunodeficiency Virus, Hepatitis B, Hepatitis C and *Treponema pallidum*. As improvements in technology have shortened the times that the assays for these antigens and antibodies take, it has become possible to offer 'same day' results for some of these tests. Where the laboratory can provide such a service, it is very helpful to the GUM clinic staff. Due to the sensitive nature of these infections, there is a chance that if a person has to return to the clinic days, or even weeks, later to collect results, they may not do so.

Genital swabs are taken to be examined in the microbiology laboratory for the presence of organisms such as *Neisseria gonorrhoeae*, *Chlamydia trachomatis*, *Gardnerella vaginalis*, *Trichomonas vaginalis* and *Candida* spp. Again, it is important to process these as soon as possible, although since overnight culture is involved for most of these organisms, it may take several days until results are available.

If an individual person does not come to a follow-up appointment, there are implications for their welfare. For example, some infections, such as chlamydia and gonorrhoea, are easily alleviated with antibiotics, but have long-term consequences if left untreated. There are also public health issues surrounding a person who has an undiagnosed and untreated sexually transmitted infection; previous partners cannot be contacted and offered treatment and future partners may be infected with a serious disease. Therefore, the agreement of and keeping to turnaround times between the laboratory and the GUM clinic is a vital contribution to individual and public sexual health.

2.5.4 Primary care

General practitioners use the pathology service a great deal, to help diagnosis where the patient's symptoms are rather non-specific and also to confirm or exclude certain diseases. For example, in a teenager whose diet is rather limited and is feeling tired, a general practitioner might send blood to haematology for a full blood count to test for anaemia and to biochemistry to test for metabolic disorders. Cervical smears are usually taken in family planning and women's health clinics or by general practitioners. Community maternity services also rely on pathology to help them care for pregnant women, particularly in the early stages, when all are routinely screened for anaemia, Rhesus and Rubella antibodies (the absence of which can be dangerous in each case) and sexually transmitted infections. Where the patient is not acutely or terminally ill, there is usually no requirement for the test result to be available within a few hours. However, like other areas of the NHS, primary care services are guided by targets to see, diagnose and treat patients within specified lengths of time. Therefore, it is important that the pathology department provides an efficient service and sends results within the agreed turnaround times, in order to assist general practitioners, district nurses and midwives in their work.

In addition to turnaround times for results, each user department will work with the pathology managers to agree which tests should be offered and the approximate number that will be sent to each laboratory per year. This is very important, as pathology must plan ahead and budget for the amount of work that it will have. The laboratory manager must be confident that they have adequate numbers of staff at each grade, the right equipment and enough money to buy containers and reagents to provide the level of service which the users require.

2.6 Evidence-based laboratory medicine

In recent years, the concept of 'evidence-based medicine' has been developed. This suggests that healthcare practitioners should make diagnoses and decisions about patients' treatment based on knowledge of up-to-date information, gained from credible research. The argument is that clinicians often prefer to rely on the way they were trained (sometimes many years previously), 'gut instinct' or a personal preference for a particular drug, rather than a critical appraisal of the relevant literature, which might challenge their way of working. 'Evidence-based laboratory medicine' (EBLM) is an extension of this, which encompasses the work of pathology departments. The idea is that front-line clinicians should use laboratory tests as part of a systematic decision-making process, while trying to reach a diagnosis and or decide on appropriate management for a particular patient. Through an awareness of published information about which laboratory tests are most useful in which circumstances, the requesting practitioner can be specific, rather than just asking for every test which they think might be relevant. The medical staff, clinical scientists and biomedical scientists in each laboratory should also contribute to the process of deciding which tests should be done in which clinical circumstances, as they have specialized knowledge of their particular area of pathology. This expert input should include data about the specificity and sensitivity of each testing method under consideration, along with awareness of the positive and negative predictive values of the test in the population served by that particular pathology service (see Chapter 6).

Laboratory tests are both time consuming and expensive, so 'evidence-based' requesting is better for both the patient and the Pathology Department. By narrowing the options, based on clinical signs and symptoms, and targeting the tests accordingly, the person asking for the tests receives an answer more quickly; the number of unnecessary investigations undertaken in pathology laboratories is also reduced. This process can also be helped by providing comprehensive relevant clinical details on the request form, which is why pathology staff try to encourage this among their colleagues on the wards and in primary care. The biomedical scientists, along with clinical scientists and medical staff in the department, should also be following EBLM in their choice of analytes to test for and assay methods, based on their knowledge of the research and technical literature.

Box 2.5 outlines the principles of evidence-based laboratory medicine, as developed by Professors Price and Christenson, who have backgrounds in clinical chemistry (Price and Christenson, 2007). Pathology test results can support patient care in three main ways, namely diagnosis, monitoring and screening. It is important to make the distinction between these three functions, since it can affect the procedure and the test analyte which is selected for the laboratory procedure.

Box 2.5   Principles of evidence-based laboratory medicine

1. Clearly define the clinical question and decide how a laboratory test will help to make the clinical decision: is the test intended for diagnostic, monitoring or screening purposes?

2. Through careful examination of published literature, select the analyte or marker to be tested for in the laboratory, on the basis of the clinical question: is this always (or almost always) outside the normal range in people with the condition in question?

3. Using results from well-designed comparative studies, choose the best test method according to technical and service delivery requirements: is it more important to use the most sensitive test or is a less sensitive, but faster assay more clinically useful?

4. Obtain information about the contribution to clinical outcome to decide whether the cost per test is justifiable: if the results of this test are available, do patients usually have a better outcome?

Based on information in Price, C. P., and Christenson, R. H. (eds) (2007) *Evidence-based Laboratory Medicine: Principles, Practice and Outcomes*, 2nd edn. Copyright © 2007, American Association for Clinical Chemistry Press

For example, if a patient presents with symptoms of a sudden onset of high fever, rigours, lethargy, nausea and vomiting, they are likely to be suffering from an acute infection and there could be a number of possible causes. If the patient has recently returned from a safari holiday which involved travelling through remote areas of a number of countries in southern Africa, malaria might be suspected. Although the patient may have reported taking anti-malarial prophylaxis, an awareness of current information would tell the requesting clinician that the most likely cause of malaria in that region is *Plasmodium falciparum* and that chloroquine-resistant strains are common. Staff in the haematology laboratory could then chose a rapid test for *P. falciparum*, using a sample of the patient's blood, which could confirm the clinical suspicion within a short time (since the testing procedure usually takes 10–15 minutes). Figure 2.2 shows the result of a rapid test for *P. falciparum*.

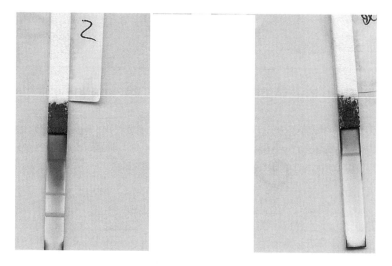

Figure 2.2   Rapid test for *Plasmodium* spp., showing positive result for *P. falciparum*. The negative control strip on the right-hand side shows a single band of colour at the top, which indicates that the test has worked as expected. The strip incubated with the patient's sample shows two further bands: the middle band, which gives a reaction when the patient sample contains parasites of *Plasmodium* spp., and the bottom band, which identifies the species as *Plasmodium falciparum*

Once the treatment had been started, blood could be collected a regular intervals (e.g. after 24 hours and thereafter daily) to monitor the levels of parasite in the patient's blood ('parasitaemia'). The best method to quantify parasitaemia is to count the malaria parasites in a thin blood film. It could be several hours before the result of this test becomes available, but it provides the degree of accuracy needed to monitor treatment. The appearance of a thin blood film containing *Plasmodium falciparum* parasites is shown in Figure 2.3.

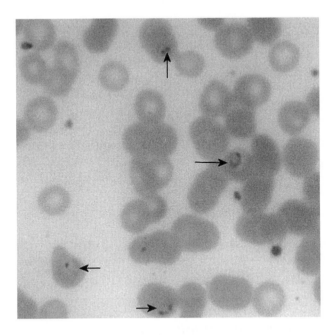

Figure 2.3   Thin blood film showing trophozoites of *Plasmodium falciparum*. The ring form trophozoites are visible inside red blood cells. Photograph courtesy of Dr A. Gunn

If the patient subsequently returned a month after completing the treatment, apparently well and showing no symptoms, it might be considered appropriate to screen their blood for the presence of parasites. In this situation, if there was any remaining *P. falciparum*, it would be expected to be found in small numbers. Hence a thick blood film, which takes longer to prepare and examine than a rapid test, but is more sensitive, would be most appropriate, as shown in Figure 2.4.

This example shows how the patient can benefit from the requesting clinician and the laboratory staff working together and applying the principles of EBLM. When users provide relevant information and make appropriate decisions about which tests to request, then the laboratory can respond with a more clinically effective service. This in turn can free up time and funding for staff training and the development or evaluation of more sensitive laboratory tests, which improves the service still further. A key element of EBLM is appraising the available evidence about the value of a particular analyte as a diagnostic marker. For individual departments, this means not only reading about studies undertaken by other practitioners, but also regular audit of their experiences with each test (see Chapter 4).

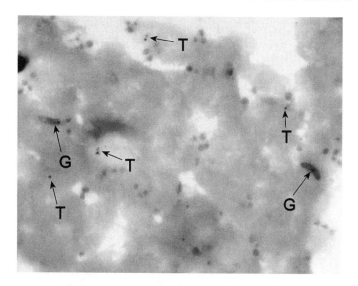

Figure 2.4 Thick blood film showing *Plasmodium falciparum* stages. Mature ring form trophozoites (T) and gametocytes (G) of characteristic 'banana-shaped' appearance are visible. Photograph courtesy of Dr A. Gunn

2.7 Improving the pathology service

Although it is difficult to collect an exact figure, NHS Pathology Departments across the UK probably receive a total of 200 million samples each year. Trusts are estimated to spend about 5% of their total budget on pathology services and, in England alone, this is thought to cost at least £2 billion per year! Data recently collected by the Healthcare Commission (Healthcare Commission, 2007) show that the number of pathology test requests is steadily increasing each year, while turnaround times are decreasing. On average, 50% of staff in a typical pathology department are biomedical scientists and their salaries account for 25% of the budget. This survey also found that in 2005, about 20% of consultants and senior staff in pathology departments were aged over 55 years and would therefore be due for retirement in less than 10 years. This will potentially cause staff shortages and a loss of specialized expertise, which will take time to replace.

In recent years, the Departments of Health across the countries of the UK have been working to 'modernize' pathology services. The aim of these initiatives is to make the service more responsive to users' and thus ultimately patients' needs. The concept of EBLM, described above, links into this objective well, as they are both focused on outcomes for the patient. 'Modernization' of the service includes the aspiration of working with users to ensure that the laboratory tests performed are those which are actually necessary and clinically useful, but also addresses the issue of where this testing work should be done.

Advances in science and technology mean that the number of different assays which are available is constantly increasingly and demand for a particular 'new' test always seems to grow once it has been introduced. The currently recommended way to meet these requirements, in a time of restricted finances and staff shortages, is for neighbouring Pathology Departments to pool resources. Large and expensive pieces of equipment, such as biochemistry analysers, which take up a lot of laboratory floor space, but which can process high numbers of specimens each day, can be purchased on behalf of all the pathology services in the group, but sited in a laboratory at just one of the hospitals in the group. Similarly, routine tissue processing could be performed in bulk at another site. Although biomedical scientists and other specialized pathology staff may have to change their normal place of work, they can gain a lot of experience in running particular assays and become specialized in certain areas. Medical consultants and biomedical scientist managers at different hospitals can help each other with clinical or logistical problems and share their ideas for good and efficient practice. This gives confidence in the quality of the test results, thus improving the service. This closer cooperation can be supported by the formal setting up of 'managed pathology networks', which have been formed in some areas of the country. This allows services to be planned more effectively and also makes it possible to establish stronger information technology links. For example, there are computer programs available which can scan stained slides being observed under a microscope by a person in one place and open them for access by another (authorized) person in a different geographical location. Within a managed network arrangement, this would allow a biomedical scientist to show a blood slide with an unusual finding to a medical consultant based at another site.

However, another aspect of responding to patients' needs is to think about their convenience. Sometimes the test method which produces the quickest result is more useful than the most sensitive one (see the malaria example above); in other situations, the simplest and least invasive procedure is most appropriate. It was been estimated that up to 40% of pathology tests are requested from primary care, rather than hospital departments. It is not common for general practitioners to collect blood samples for pathology tests during routine appointments; the doctor usually issues a request form and the patient has to attend a phlebotomy session either at the surgery at another time or at a hospital. Similarly, patients requiring urine tests often have to take a collecting pot away with them and deliver the sample to the surgery or the hospital themselves later. In many of these cases, the patient is not acutely or terminally ill and it can be inconvenient to take more time off work or reschedule arrangements for the day, to return to the general practice or go to the hospital laboratory – which could be some distance away, due to centralization of services! Sometimes, when a patient was reluctant to have the test in the first place, they may not return to have the sample taken or to collect the results. For example, many people at risk from high blood cholesterol levels feel perfectly well and consider themselves to be healthy. When their general practitioner suggests a cholesterol screen, the patient might consider it unnecessary. If they are asked to attend the hospital, a week later, to have a venous blood sample collected in order to have the cholesterol test, that patient may not consider it important enough to make the time to go. Thus the opportunity to screen someone for blood cholesterol levels and offer those with high levels advice about their diet and lifestyle may have been missed. This situation could be avoided if the patient could have the test done at the surgery at the time of the initial visit to the doctor and be given the opportunity to discuss the results during the consultation. This is 'point of care testing', which is set to become an increasingly important part of the provision of pathology services in future.

2.8 Point of care testing (POCT)

Diagnostic pathology tests which are performed outside the main laboratory are variously referred to as 'point of care' tests, 'bedside tests' or 'near patient tests'. Drinking a patient's urine to test for sweetness, as an indication of diabetes mellitus (described in Chapter 1), is the original 'point of care test'. During the 20th century, the tendency was for testing to be moved from the wards and into the laboratory, due to the need for complicated equipment and expensive reagents. However, technological advances from the 1980s onwards meant that everything, from computers and televisions to mobile telephones and gadgets for listening to recorded music, have become smaller and work faster. Companies making diagnostic tests for pathology laboratories followed this trend, by developing portable analysers, 'hand-held devices' and tests which require little or no training to carry out. The scientific reaction taking place inside the device is often complex, but it is not necessary for the person doing the test to understand it in order to achieve a meaningful result. A good example of this is the pregnancy test kits which can be bought 'over the counter' in high-street pharmacies and supermarkets for use at home. The kits are presented in a 'user-friendly' way, with simple operating instructions. The testing device has an inbuilt control, which enhances the reliability of the result, and this is the generally accepted way for UK women to determine whether they are pregnant (see Figure 2.5). Indeed, the 'home testing' devices are now so sensitive and reliable that they are often used by biomedical scientists to test the samples sent by general practitioners to pathology laboratories!

There are a wide range of point of care tests available and the list is increasing steadily. One of the simplest tests is the urine dipstick, which is the successor of the 'drinking urine' method. This comprises a cellulose strip embedded with reagents which detect certain components of urine, through a colour change,

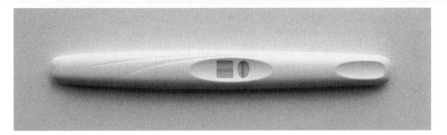

Figure 2.5 'Point of care' pregnancy testing device, showing a negative result. The oval window on the right-hand side of the assay strip area is the control panel; the vertical line indicates that the test has worked as expected. The square window on the left-hand side of the assay strip area is the test panel; the single horizontal line indicates that the patient is not pregnant (in a positive result, there would also be a vertical line in this square window, forming a cross)

which can be read by eye. The most commonly used type can show the presence of a range of substances which could indicate disease when found in urine, such as red blood cells, white blood cells, ketones, nitrites and protein (see Figure 2.6). The result is essentially qualitative, but it can give a good indication of certain abnormalities – for example, an unusually high concentration of protein and white blood cells can be an indication of a urinary tract infection.

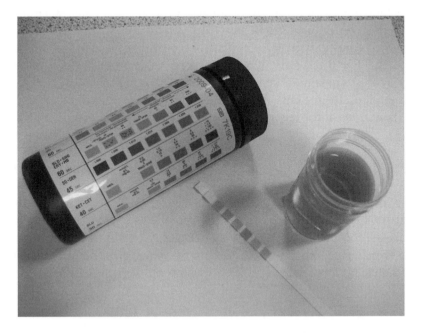

Figure 2.6 Urine dipstick test strip showing result. Concentrations of specific chemical and cellular components of urine are estimated by comparing colour changes on the test strip with the indicator chart on the bottle. Photograph courtesy of Miss Lynnsay Fuller

Kits which use a cellulose strip as the solid phase in an immunoassay for a specific antigen or antibody have also been devised. The urine pregnancy test is an example of this (see Figure 2.5). There are also tests for infections such as malaria (Figure 2.2) and HIV in blood samples or respiratory syncytial virus and influenza virus in respiratory specimens; there are also POCT kits to detect drugs of abuse in blood and urine samples. Most of these have an initial simple preparation step to release the antigen or antibody of interest from the surrounding cellular material and also require the addition of reagents during the assay. However, they still provide rapid results. The result is a qualitative colour change on the strip which is read by eye.

Tests involving the use of a hand-held reader were originally developed for blood glucose monitoring and they are widely used, particularly by diabetic patients to help monitor their blood glucose concentration (glycaemia). The test requires a drop of whole blood from a finger prick, which is placed on a test strip embedded with a reagent which changes colour to indicate blood glucose. The intensity of the colour change is proportional to the concentration of glucose. The strip is read by a hand-held device, which gives a quantitative result. Using this reading, the patient (or healthcare professional treating them, if they are acutely ill) can decide whether it is necessary to adjust their insulin dose or administer glucose. This technology has been adapted for measuring other biochemical analytes such as blood cholesterol.

Benchtop analysers for the measurement of blood gases and other analytes are commonly used in A&E Departments (where a rapid answer is required) and critical care units (where the patient's condition needs to be constantly monitored). For these tests, blood has to be taken from the brachial artery into a syringe containing heparin and mixed thoroughly and rapidly placed in the machine for reading. A quantitative result is produced, which can give clear indications about the nature and seriousness of the illness in a particular patient. Hand-held devices are now becoming available which require only a few drops of blood on a testing cassette strip; these can test for blood gases and other biochemical and haematological parameters such as cardiac markers and haemoglobin.

Non-invasive methods are also used, such as the breathalyser test used to test for alcohol levels, the urea breath test for *Helicobacter pylori* and the sweat test for high salt concentrations, which may indicate cystic fibrosis. Since no sample has to be collected, this type of test is simpler to perform and more acceptable to the patient, so it is a thriving area of research, but at present the scope is fairly limited.

2.9 Role of POCT in patient care

POCT has the potential to be of great clinical value. Most POCT procedures produce a result more quickly than the equivalent test performed in the main laboratory, which means that staff treating the patient do not have to wait as long to use it in their decisions about managing the illness. In addition, methods are usually made simple enough that anyone who has received some basic training can carry them out. Another possible advantage is that the type of sample needed is usually less invasive than the equivalent main laboratory test (for example, a drop of blood from a finger prick rather than a venous sample), which makes it more acceptable to the patient. The logistics of transporting the specimen to the laboratory are no longer of concern and the likelihood of a mistake in assigning the correct patient details to the sample should also be reduced. All this means that healthcare professionals without any specialized knowledge or training in biomedical science can use POCT devices – for example, doctors on wards and in general practice, nurses, pharmacists, allied health professionals – as can patients.

However, there are some disadvantages to using POCT. In condensing complicated technology into a small hand-held or desktop device, the manufacturers have invested a lot of time and expertise, so the equipment and reagents are usually comparatively costly. Reagents have a limited shelf-life and, if the test is not done often, these may need replacing when only a small amount has been used, which rarely happens for routine procedures in main pathology laboratories. Hence the cost per test is usually considerably greater for POCTs than the equivalent main laboratory test. Although the technology is becoming very sophisticated, the POCT is usually less sensitive and specific than the best laboratory method. The requirements in the clinical situation must therefore be carefully balanced by the consequences of false-positive or -negative results. The positive and negative predictive value of each test in a given population should also be taken into account (see Section 2.6 above and Chapter 6).

From the biomedical scientist's point of view, another problem is the issue of quality assurance. Anyone using a POCT kit must be trained before using it, and this should include instructions on performing regular quality control, to ensure that the results are reliable. However, in a busy and pressured clinical situation, this tends to be seen as an 'added extra' rather than a necessity and it can be neglected. This means that systematic errors can creep in unnoticed and even random errors could be ignored (see Chapter 4) if the device is working sufficiently well to give a result. The system for recording and collating results can also be an area of uncertainty; the practitioner who performs the test can record the result in the patient's notes, but there should be a central record of all the results. Regular audits of each POCT need to be carried out to ensure that it is providing the best information for patient care, according to the ideas of EBLM discussed above. Biomedical scientists are trained and experienced in all these matters and so it is appropriate for them to be closely involved in the quality assurance of any POCT devices used in any hospital departments or primary care centres. Opportunities to specialize in POCT are opening up in a range of settings for a number of groups of healthcare practitioners.

For example, community pharmacists and nurses working in general practice surgeries are being encouraged to offer tests such as blood cholesterol levels, for their customers or patients, respectively, as part of providing advice on a healthy life style. There is no particular reason why biomedical scientists should not have direct patient contact and perform these tests, and also operate the quality assurance system. The IBMS offers a Certificate in Primary Care to support members who want to develop their careers by working away from the main laboratory. It is important to see the tests done in the central Pathology Department and POCT performed outside the laboratory as complementary. For urgent tests and primary care, the POCT option may be the most appropriate, but the test carried out in the main laboratory may be more scientifically accurate and cost effective. As the example of the malaria diagnosis above illustrates, in some situations a rapid answer is required, whereas in others, the expert judgement of a specialized biomedical scientist, working in a well-equipped laboratory, is essential to achieve the best result for the patient.

2.10 Conclusion

This chapter has shown how the various staff groups within pathology can work together to provide an appropriate and efficient service for the users and ultimately the patient. Biomedical scientists need to be aware of the role of pathology in patient care, to be responsive to the requirements of the users and also to apply advances in science and technology to the diagnostic context, in order to improve the service.

Quick quiz

1. Name the eight main groups of staff who work in a Pathology department.

2. Name three specific jobs which are performed by associate or assistant practitioners.

3. What qualifications are required to work as a cytoscreener?

4. In which ways does the role of a clinical scientists differ to that of a medical practitioner within pathology?

5. What proportion of diagnoses rely on pathology test results?

6. What is meant by the 'turnaround time' for a test result?

7. Name five users of a typical pathology service.

8. For one of these users named in question 7, describe why an efficient pathology service is important for them to work effectively.

9. Describe the four principles of evidence-based laboratory medicine.

10. What proportion of pathology test requests are estimated to come from primary care?

11. What is meant by 'point of care testing'(POCT)?

12. List two advantages and two disadvantages of POCT.

Suggested exercises

1. Draw a diagram showing the line of accountability for all grades of staff in your department. For each job, describe their main duties, responsibilities and who the post-holder would supervise. Also indicate who deals with problems or queries from staff in each group.

2. Investigate the mix of grades of staff in a department which operates differently to yours (for example, one with more assistant practitioners than yours or one with clinical scientists). Write a reflective essay on the reasons for the differences in staffing between your department and the other one and consider whether the mix of grades in your laboratory is appropriate.

3. Select a POCT device, investigate the scientific basis for it, the technical quality of the results and the test procedure. Research the literature for studies which have evaluated the test against existing laboratory tests. Prepare a report about this test for a mixed group of healthcare practitioners who are considering implementing this test on their ward, in their clinic or in their primary care service, clearly indicating whether or not you would recommend that they use it, giving reasons.

Suggested references

Carter, Lord of Coles (2006). *Report of the Review of NHS Pathology Services in England*. London: Department of Health.

Department of Health (2005). *Modernising Pathology: Building a Service Responsive to Patients*. London: Department of Health.

Healthcare Commission (2007). *Getting Results: Pathology Services in Acute and Specialist Trusts*. London: Commission for Healthcare and Audit Inspection.

Price, C. P., and Christenson, R. H. (eds) (2007). *Evidence-based Laboratory Medicine: Principles, Practice and Outcomes*, 2nd edn. Washington, DC: American Association for Clinical Chemistry Press.

Price, C. P., St John, A., and Hicks, J. M. (eds) (2004). *Point of Care Testing*, 2nd edn. Washington, DC: American Association for Clinical Chemistry Press.

3

Communication for biomedical scientists

3.1 Introduction

Biomedical scientists need to be able to communicate, as both scientists and healthcare professionals, with a wide variety of people. They have to use the specialized technical terms, abbreviations and jargon employed in a particular field, both when interacting with other scientists informally and when presenting results from their research in the public domain. As healthcare professionals, biomedical scientists must be also capable of communicating effectively with colleagues within the Pathology department, in addition to liaising with other staff in the hospital and in primary care. They must be able to explain practical requirements (such as arrangements for specimen collection) and technical issues (for example, why the turnaround times for some tests are longer than for others), without using shorthand which only other biomedical scientists in their department would know. Biomedical scientists must also be able to adapt their language for a variety of other groups and situations. There are increasing opportunities to have direct contact with patients and, in this situation, they must avoid scientific and medical terms, which may confuse and in some cases distress the patient. As part of their professional development, biomedical scientists may have opportunities to disseminate aspects of their professional practice to audiences beyond their own 'peer' community of biomedical scientists, for example to other healthcare professionals, patient interest groups and the general public; again, there are appropriate formats and language for these types of communication, which are used to ensure that the information is clearly understood. This chapter will consider some principles of communication for biomedical scientists and how they can apply them effectively in their capacities both as scientists and as healthcare professionals.

An Introduction to Biomedical Science in Professional and Clinical Practice   Sarah J. Pitt and James M. Cunningham
© 2009 John Wiley & Sons, Ltd

3.2 Communicating as a scientist

Scientists communicate with others for a variety of reasons, such as presentation of ideas and theories, description of experimental methods, interpretation of results or review of the available literature on a particular topic. It is important to be able to exchange information and ideas with a range of people, including experts working in the same field of study, specialists in other disciplines and other professional groups for whom an understanding of recent scientific advances may be relevant. Scientists' research projects can vary considerably in complexity and duration and often there is no neat predictable 'end' to work towards before communicating the results. It can be very useful for scientists to present research-related findings at different stages of a project, for example, presenting preliminary data or even before this at the 'big idea' stage of formulating a hypothesis and before collection of any data. Feedback from fellow scientists at these earlier stages of the project can help you to assess the validity of your approach and methodology and perhaps make you aware of relevant findings from other projects not yet published. Apart from communicating to other scientists, communications may be aimed at groups of people who do not have a specialist background, for example patient groups, the general public and family and friends. Although trying to explain a scientific principle to a non-scientist can be hard, it is a good test of your own understanding; expressing thoughts and ideas in an informal exchange, written or oral, can sometimes be a great help towards clarifying or rationalizing complex ideas or thought processes.

3.3 Communicating as a healthcare professional

In the modern healthcare service, biomedical scientists rarely spend their whole working day performing laboratory tests in isolation. Within the laboratory, biomedical scientists need to discuss technical problems or scientific findings with colleagues and pass on results to more senior biomedical scientists or medical staff, to allow them to be communicated appropriately to the requesting clinician (Figure 3.1). Advances in technology, such as the automation of serological tests, mean that laboratory equipment is often designed to perform tests across several pathology disciplines. This means that biomedical scientists are increasingly being required to operate as multidisciplinary practitioners and to teach non-specialists about their particular area of expertise. As healthcare professionals, biomedical scientists must always communicate clearly, demonstrating understanding and respect for the role of the person with whom they are interacting. All staff within pathology, whether they are assistant practitioners, healthcare scientists, managers, nurses, medically qualified staff or administrative staff, contribute to the provision of a high-quality diagnostic

Figure 3.1 A biomedical scientist discusses a set of results with a senior colleague

service. The same professional attitude must be shown when communicating with colleagues from outside the Pathology department, including general practitioners, hospital-based doctors, nurses, other allied health professionals and staff in the occupational health, human resources and finance departments. It is important to bear in mind that colleagues in other departments gain an impression of the reliability and helpfulness of the Pathology department through their interactions with individual biomedical scientists; since most communication is via letter, e-mail or the telephone, which can be impersonal, it is important to take care over this and follow local protocols and standard operating procedures (SOPs), as appropriate. Opportunities for biomedical scientists to work outside the confines of the Pathology department are starting to open up. Some biomedical scientists are working in primary care, using screening tests in general practitioners' surgeries, while others are taking on responsibility for coordinating the 'point of care' testing within hospitals. In both of these examples, the biomedical scientist would be working with staff who were not familiar with pathology and would also have more contact with patients than their laboratory-based colleagues. Posts such as these bring the opportunity to raise the profile of biomedical science profession and the chance to develop roles for biomedical scientists, as part of multiprofessional teams, outside the Pathology department. The professional training and scientific expertise that that biomedical scientists have mean that they can make a very important contribution to patient care (for example, in emphasizing quality assurance – see Chapter 4).

Since July 2006, it has been a legal requirement for HPC registered practitioners to participate in continuing professional development (CPD) activities in order to keep up to date in their practice (see Chapter 8). Attending and presenting posters and talks at conferences is a key element of CPD. It is often not possible for all staff in one department to attend the same conference, so it is important for those who go to share the information and insights gained on their return. One way of achieving this is giving a short talk – for example, as a lunchtime meeting. Another possible way of carrying out CPD is to give an oral presentation, based on a scientific or professional article, to the rest of the department. Both types of talks are examples of biomedical scientists communicating as professionals and should be planned carefully. They also go towards fulfilling both the HPC requirement to undertake CPD and the IBMS Code of Conduct point 6: 'Strive to maintain, improve and update their professional knowledge and skill'.

3.4 Delivering the message: applying principles of effective communication

It is important to remember that the most effective communication is a two-way process between the person who has a message for others to hear and the intended audience. Being able to do this well is a fundamental part of being a healthcare scientist, since the work carried out in the laboratory may have a bearing on many other people, albeit sometimes indirectly. There are a number of possible ways to exchange information in a professional situation and it is important to consider the context and select the most suitable means of addressing your audience. Taking care over this shows that you take the topic seriously and respect the people with whom you are communicating. Conversely, not doing so can distract from the message you are trying to give. For instance, if you hand in an essay for assessment, which is handwritten, in pencil, on a scruffy piece of paper and is full of inaccuracies and spelling mistakes, the reader is likely to assume that you have not thought deeply about the topic – even if you have. In this situation, you indicate to your reader that you are not really interested in the ideas you have presented, which can detract from the particular scientific or professional points which you were trying to convey. To maximize the effectiveness of any communication, seven principles for effective communication are suggested (Box 3.1), and these will be discussed below and set in the particular context of communicating as a scientist and a healthcare professional. In addition to common sense and good scientific practice, it is useful also to bear the biomedical scientists' Codes of Conduct in mind when interacting with colleagues, other healthcare professionals and patients. They will be discussed further in this section, in the light of the HPC

Standards of Conduct, Performance and Ethics (HPC SCPE) and the IBMS Code of Professional Conduct (IBMS CPC) (see Chapter 1).

Box 3.1   Principles of effective communication

- Choose the most appropriate means of communication.
- Be clear about the main message to be understood and remembered.
- Consider the audience.
- Work within established protocols and guidelines.
- Follow scientific conventions.
- Be accurate and honest.
- Listen carefully.

3.4.1   Choose the most appropriate means of communication

There are a number of possible ways to exchange information in a professional situation, and these can be categorized broadly as either 'informal' or 'formal' and then as 'oral' or 'written' forms of communication. Informal communication can be appropriate for new ideas or speculative interpretation of results which are not intended as a permanent record or publication; this can take place by conversations, e-mail and meetings that are held between small groups of individuals. Formal means of communication include presenting results from a research project at a professional meeting, for example by means of a talk or poster, participation in a scheduled meeting (i.e. one which has a prearranged time and venue, an agenda and recorded minutes), writing a report or article for publication or submitting a grant application to obtain funds to support a research project.

How do you go about choosing the most appropriate means of communication? The best medium of communication to reach the target group may not be obvious. In addition to communicating your research findings to scientists in the same field, there may be important implications of your work for other healthcare professionals (for example, the validation of a new diagnostic test may be of importance to both medical staff and patient groups). Should you communicate that information by talking face to face, telephoning, sending an e-mail, writing a letter, holding a formal meeting, giving a short talk, writing an article or presenting a poster or oral paper at a conference? For example, you may have recently completed a research project and you want to make available a full account of this to scientists in the same specialist area of research. The most appropriate medium in this case is usually a full-length written article in a specialist journal that you know is read by your fellow researchers working in the same field. If you are not sure of the best approach, Table 3.1 may be helpful as a guide to suggest modes of communication of different 'messages' to different

Table 3.1 Suggested methods of communicating particular messages

Message	Suggested ways of communication	Comments
Practical information exchange, e.g. confirming time, date or venue of a meeting	Talk to the person, e-mail or telephone call	This is an informal communication, no understanding of concepts or help with tasks required so can be brief, but not brusque!
Requesting help or advice	Talk to the person, e-mail or telephone call	This may be informal if the person is an immediate work colleague or a friend; however, you are likely to have a more satisfactory outcome with someone you do not know well if you start formally, e.g. by addressing the person as Mrs or Dr
Guidance on service provision	Leaflet, talk, webpage	This is formal and should be carefully structured so that users are given the essential information and find it easy to follow. It is therefore important to consider the background of your target audience at the planning stage
Preliminary data from a research project	Poster, short talk	This may be formal and both these methods are satisfactory. Which one you choose may depend on the opportunities – for example, you may submit an abstract to do short talk at a scientific meeting, but be invited to do a poster instead
Outcome of an internal audit, e.g. quality, safety	Poster, written report, laboratory meeting	This should be presented in a formal way. It is a good idea to present findings orally in a laboratory meeting, but you should also have them in an easily accessible written format too for people who were unable to attend the talk
Final outcome of a research project	Journal article, short talk	This is a formal communication and, if you have the opportunity, do both journal article and oral presentation at a conference. In each case, follow the prescribed format, which improves the chance that readers and hearers will understand the key points of your findings

target groups (local, national or international community of peer-scientists or others).

3.4.2 Be clear about the main message to be understood and remembered

Think carefully about exactly what the message is and the best way for this to be communicated with clarity and brevity, so that the salient points stand out and cannot be misunderstood. While planning something to reach a wide audience, for example a talk or an information leaflet, it is worth checking that you have kept to the original core message(s) before producing the final version. Working with this in mind will ensure that you meet HPC SCPE 7: 'You must always maintain proper and effective communications with patients, clients, users, carers and professionals', and IBMS CPC 1: 'All members shall always exercise their professional judgement, skill and care to the best of their ability'.

3.4.3 Consider the audience

Use a style of presentation and language which will maximize the engagement of the readers or listeners. Think about which terms would be familiar to an audience of scientists in your discipline, pathology staff from other disciplines, other healthcare professionals, people outside hospital, patients or a mixture of these. Use technical terms carefully and only when they are appropriate to the readers of your article or the audience at an oral presentation. Always define abbreviations and avoid unnecessary 'jargon' that may not be understood by the target audience. For an oral presentation, the most appropriate style and language will also vary according to the size of the audience (one person, a small group, a large group). For example, there will a great difference in the information you provide and the way you present it between giving a talk to school children about the biomedical science profession and informing hospital-based pathology service users of a change in your department's provision. Always remember that you are a representative of the biomedical scientist profession and be clear and polite at all times. By catering for the audience effectively, you will be following HPC SCPE 7: 'You must always maintain proper and effective communications with patients, clients, users, carers and professionals', and IBMS CPC 5: 'All members shall always act in good faith towards those with whom they stand in professional relationship and conduct themselves so as to uphold the reputation of their profession'.

3.4.4 Work within established protocols and guidelines

These are in place to make communication clearer, so, for example, make sure that you keep to the allotted time when giving a talk, follow the journal guidelines when submitting a paper or know the correct way to format a letter. In the clinical setting, these are also in place to make sure that all staff provide the relevant information, to the appropriate person, and stay within the limits of their professional practice. Hence you should always follow standard operating procedures about communication within the laboratory. In particular, you should be familiar with the system for reporting test results in your department. In this way, you would be fulfilling the requirements of HPC SCPE 3: 'You must always maintain high standards of personal conduct', HPC SCPE 6: 'You must always act within the limits of your knowledge, skills and experience and if necessary, refer on to other professionals', and HPC SCPE 13: 'You must always carry out your duties in a professional and ethical way'. You will also meet IBMS CPC 1: 'All members shall always exercise their professional judgement, skill and care to the best of their ability'.

3.4.5 Follow scientific conventions

These exist to reduce ambiguity and misunderstanding when scientists are exchanging information. Abbreviations should always be defined and SI units and referencing systems should always be used (see Box 3.2). Among other aspects of the codes of conduct, you will be keeping HPC SCPE 13: 'You must always carry out your duties in a professional and ethical way', IBMS CPC 1: 'All members shall always exercise their professional judgement, skill and care to the best of their ability', and IBMS CPC 2 'All members shall always fulfil their professional role with integrity, refraining from its misuse to the detriment of patients, employers or professional colleagues'.

Box 3.2   Conventions of scientific communication

The conventions which scientists use when communicating with each other can sometimes seem unnecessarily pedantic. However, they are actually an invaluable way of avoiding confusion and ensuring that colleagues can understand your work.

For instance, the *Système International d'Unités* (SI) for defining units is used throughout the world. It was introduced in the 1960s by the *Conférence Générale de Poids et Mesures* (General Conference on Weights and Measures) in order to standardize units which scientists used when

---

describing their methods, for example in peer-reviewed journal articles. The metric system was introduced in France after the French Revolution, but the United Kingdom and other countries used the imperial system of measurement until the late 20th century. It is complicated and inexact to convert weights, volumes and lengths from metric to imperial. One ounce of a substance is calculated as 28.35 grams; 1 millimetre in length is equivalent to 0.3937 inches. In scientific experiments which use small amounts of materials, it is clear how inaccuracies could creep in when a scientist who used the imperial system tried to repeat an experiment conducted by a colleague using metric measurements. Thus, metric measurements are always quoted in published methodologies. Another point to note is that there is often an accepted unit of measurement for particular parameters. For example, the reference range for white blood cell count is usually given as $4.0–11.0 \times 10^9/L$; a busy doctor is likely to look at the beginning of that result and check that the number lies between 4 and 11. If the laboratory changed its style and issued a white blood cell report of $0.005 \times 10^9/mL$ for a patient, this could cause unnecessary distress to the doctor and patient, despite being a white blood cell count within the normal range. Thus, by following the agreed convention for units, an individual biomedical scientist can be confident that others will clearly understand their results.

Similarly, the International Committee on Zoological Nomenclature (ICZN) agreed to use certain rules when naming animals and plants at the turn of the 20th century. Since at that time scientists tended also to be well versed in Classical Languages, the names they chose were often based on Greek words or had Latin-type endings. By convention in English grammar, 'foreign' words or phrases are written in italics; therefore, it was agreed that these 'proper' biological names would be italicized. In biomedical science, this rule in particularly important in microbiology. Consider the following sentence: 'In the laboratory diagnosis of leishmaniasis, thin blood films can be stained with Leishman's stain and examined for the presence of *Leishmania* spp.' A microbiologist will expect that other scientists will follow the accepted convention and so would immediately see which word is the name of the parasite and which refers to the type of stain. Following this rule also ensures that potential confusion such as that between the name of the bacterium *Haemophilus* spp. and the condition haemophilia is easily avoided.

Scientific advances are usually achieved when scientists know the background in their chosen field. Thus, 'new' methods and ideas are generally based on earlier work and previous authors must therefore always be credited. This acknowledges the other peoples' work, which is professional

---

Box 3.2   (Continued)

and courteous, and also allows the reader or hearer to see the roots of your work. For example, the paper by Saiki *et al.* (1988) mentioned in Box 3.3 ('peer review'), which describes the method for the polymerase chain reaction, has been cited in over 14 000 subsequent papers. The required format for citing references varies between academic institutions and journals. It is therefore important to ascertain the correct style before submitting work.

3.4.6   Be accurate and honest

Scientists should be objective at all times when describing and explaining their findings. The scientific process relies on accurate and honest reporting of the experimental process and data obtained; omitting results or inventing data, however minor, is a serious breach of an expected 'code of conduct' that all scientists should follow. Similarly, an individual presenting another person's work as their own, in order to gain advancement, is plagiarism, which is unacceptable in the scientific profession, as elsewhere. As healthcare professionals, biomedical scientists must try to be objective at all times about their professional attitude and their laboratory findings. If you have made a mistake during a test, it is better to be honest about it, as it may affect the result, which may compromise patient safety; equally, an error in laboratory procedure may create a situation where the safety of staff within the department is compromised. Everyone makes mistakes at some time, so if you tell someone the truth as soon as you are aware of the problem, it makes it much easier to deal with and rectify. This is therefore part of being a good colleague and team member. Similarly, it is more helpful to users of the pathology service if staff are truthful about turnaround times for tests and take responsibility for any mistakes that the laboratory has made. This actually makes the department seem more trustworthy in the long run and helps maintain a good relationship with users. Keeping to this ensures that you comply with HPC SCPE 7: 'You must always keep accurate patient, client and user records', HPC SCPE 14: 'You must always behave with integrity and honesty', IBMS CPC 2: 'All members shall always fulfil their professional role with integrity, refraining from its misuse to the detriment of patients, employers or professional colleagues', IBMS CPC 3: 'All members shall always seek to safeguard patients and others, particularly in relation to health and safety', and IBMS CPC 5: 'All members shall always act in good faith towards those with whom they stand in professional relationship and conduct themselves so as to uphold the reputation of their profession'.

3.4.7 Listen carefully

Scientists should keep an open mind at all times, which includes being respectful of others' perspectives and points of view. It is important to listen to the person or people with whom you are communicating, to ensure that you understand the information that they are conveying to you. Considering other people's ideas or perspective against yours can be an extremely useful way of gaining insights into your own understanding of concepts or interpretation of results. One form of feedback that can be extremely useful arises from the 'peer review' system used in most forms of scientific publication. This is where the article would be reviewed by one or more experts in the field who would check the data presented from results, test any conclusions which are made about their findings with academic rigour and would recommend to the journal Editor whether the article is acceptable for publication. This process is usually done anonymously, which is intended to allow the reviewers to be honest. Once an article has been accepted and published, then other scientists can read it, try out the methods for themselves and in turn communicate their findings with the scientific community. This process is an important part of the process of scientific communication as it ensures that published articles have withstood a level of expert scrutiny regarding the integrity of the study and the information provided. Giving other scientists greater confidence in the quality of a publication enhances the impact of new ideas and findings, and an example of a peer-reviewed paper that has had considerable importance across many scientific and medical fields is given in Box 3.3.

Box 3.3   The polymerase chain reaction method: an example of a high-impact peer-reviewed scientific publication

The polymerase chain reaction (PCR) is a widely used method in biomedical sciences for the detection and amplification of specific fragments of nucleic acid. The principles were outlined in peer-reviewed research articles published in the 1970s, which describe a method which requires heating the DNA of interest, to denature it, thus allowing a DNA polymerase enzyme to anneal and copy the original piece of DNA. Each time the test mixture was heated, the enzyme was denatured and had to be replaced. When a team of scientists in the United States refined the method by using a thermostable DNA polymerase enzyme (Saiki *et al.*, 1988), it was useful to other researchers. The first applications of the method were in the detection of genetic abnormalities in haematological disorders of humans, but as scientists in different fields read peer-reviewed articles about this method and tried it in their own areas of interest, it

---

Box 3.3   (Continued)

become apparent that the technique is very useful to detect small amounts of nucleic acid in many circumstances. For example, PCR can be used to identify pathogenic microorganisms in clinical samples and also tumour markers in human cells. For a recent review of PCR methods, see Theaker (2007) and for clinical applications see Lo *et al.* (2006).

When communicating with people in a professional capacity, it is important to make sure that you respond to their problems and queries with accurate and helpful information. This requires careful listening. You must also be careful about how you deal with the content of any exchange, particularly if it is of a sensitive or personal nature, whether it is about a colleague or a patient. The codes of conduct relating to this are HPC SCPE 2: 'You must respect the confidentiality of your patients, clients and users', and IBMS CPC 4: 'All members shall always treat with discretion all confidential and other information requiring protection and avoid disclosure to any unauthorized person the result of any investigation or other information of a personal or confidential nature gained in the practice of their profession'.

3.5 Communication techniques

Regardless of whom you plan to communicate with and how you intend to do so, it is important to prepare well. This helps to make sure that the person or people with whom you are communicating understand the points you want to make and also improves the outcome for you. For example, if you are asking a colleague or a tutor for help, you should clearly explain your problem and, if you have a number of issues to discuss, it can be helpful to write down a 'check list' for yourself. They are more likely to give your items the attention you would like if you choose a time when they are not too busy and are polite. This example illustrates the point that in any form of communication, the aim is to engage the audience gently but firmly.

3.5.1 Informal communication

Perhaps the most informal method of communicating is talking through problems or ideas with a person you encounter in a chance meeting – for example, in a corridor! Often this will lead to arranging a more convenient time to discuss the issues. A telephone call to someone you know, to ask them a quick question, can also be informal. It is very easy to exchange information and ideas with someone informally via e-mail. However, it is essential to remember that in a professional context, the use of casual language, slang or 'text speak' is not appropriate, even in an informal exchange, and can in some cases offend the other person.

Use of e-mail to contact people is ubiquitous in the modern workplace. It is a useful medium, as it has the potential to allow the sender to compose their thoughts, while still being relatively informal and immediate. There is also the advantage that the person receiving the e-mail can read and respond to it in their own time, when sometimes a telephone call interrupts them at an inconvenient moment. However, because writing and sending e-mails are so straightforward, it is easy to forget basic conventions of communication as a professional. Box 3.4 gives some tips on what to do and what to avoid in e-mails to colleagues and people with whom you have to liaise at work.

Box 3.4   Tips for writing e-mails in a professional capacity

- Give your message a title, particularly if the person you are sending the e-mail to is not expecting a message or does not know you very well; they may think it is 'spam' and delete it!

- Start your message with a suitable form of address, such as 'Dear Mr Smith', or 'Hi Jane' if you know the person; it is not good practice to use more familiar forms with someone you have never met or to just start the message with the person's name, as both of these could be taken as rudeness by the receiver.

- Use correct grammar and spelling in your message and avoid the codes you use in text messages to friends, as this can lead to confusion.

- Do not use a lower-case 'i' to denote the personal pronoun 'I'; this can give the impression of both rudeness and poor education.

- DO NOT WRITE THE WHOLE MESSAGE IN CAPITAL LETTERS, as this can be taken to stand for shouting and would be interpreted as a rude and aggressive attitude.

- Explain the reason for your message clearly and politely.

Box 3.4   (Continued)

- If the explanation starts getting complicated and your text takes up a whole computer screen, consider talking to or telephoning the person instead.

- Use a suitable form of words to end your message, such as 'Kind regards' or 'Best wishes'; 'Cheers' is rarely appropriate in a professional context and 'CUl8ter' is never acceptable.

3.5.2 Formal communication

When approaching the three main forms of formal communication for biomedical scientists – posters, talks and journal articles – the fundamental principles are the same. You need to catch the audience's attention from the beginning, to persuade them that it is worth spending a few minutes (or an hour) of their life listening to or reading about the information you want to present. However, if you are too garish about it (i.e. have very loud colours on a poster or talk aggressively), then your audience may become distracted or even annoyed. On the other hand, your delivery must not be too bland or boring, because that conveys the impression that you think the subject is boring, and again the audience will lose interest.

3.5.3 Posters

A popular method of communicating information at scientific meetings is by a poster presentation. This is a good way to start learning how to present at conferences; indeed, some data lend themselves better to a poster than a talk. At some meetings, ordinary delegates are not invited to apply to present short talks, but they can submit posters. On a poster, the description of the research study is displayed on a poster-board – usually A0 (841 × 1189 mm) or A1 (594 × 841 mm) – so that it can be read comfortably by small groups of people. Posters are usually displayed simultaneously by a number of authors at set times during a scientific meeting and meeting participants can both read the poster displays and discuss the work in an informal way with the authors. Unlike full scientific papers, the poster design is often not prescribed, other than the constraint of the poster dimensions. It is common for authors of oral and poster presentations to be asked to submit short abstracts of around 250 words for publication in the proceedings of the scientific meeting.

A poster is intended to be a visual experience for the reader, so it should be easy for them to read it and to follow the message you are intending to convey. It is therefore important to plan the layout carefully. It is best to choose a simple, plain background, made up of one primary colour or perhaps two blended, similar colours. An elaborate pattern or picture is distracting for the reader's eyes and a very dark or light background can make it hard to read the writing. A poster usually comprises a mixture of text and visual representations of data, such as tables, graphs and diagrams. For the main body of text, it is essential to select a font size that is large enough to read comfortably at a distance of a few feet (about a metre) and of a simple type face rather than an elaborate one. Thus, Arial or Times New Roman at a font size between 30 and 40 point would be suitable. One of the fonts which resembles handwriting, although attractive, can be very hard work to decipher on a poster; similarly, if you underestimate the size of the finished product and use the font size of 10 or 12 point which you usually use for A4 documents, the reader will struggle to read the text. In either case, if a poster requires a lot of effort to engage with, your potential audience will probably pass it by without taking in the information you would like them to understand. It is a good idea to include some colour in a poster, for example using a coloured font for headings. However, be careful to choose colours which contrast well with the background; primary colours are usually best and shades of yellow and light greens should be avoided altogether. Summarizing information or data in diagrams, tables or graphs is always useful, if done well. For the writer, it can save complicated explanations in the text and for the reader it can help to understand the message quickly. However, they are only helpful if they are given clear, explanatory headings and are labelled properly. Tables should be numbered sequentially, i.e. Table 1, Table 2, . . ., and given descriptive titles and have suitably labelled columns. Graphs and diagrams should be numbered as figures, i.e. Figure 1, Figure 2, . . ., and again have captions that explain the main point of the visual aid. Graphs must have clearly labelled axes and diagrams should have annotations pointing out the salient features.

Posters usually have an introduction to the topic under consideration, the main point of the presentation (for example, the methods used and data collected in a piece of research or an important new safety message) and some sort of summary or conclusion at the end. How these are arranged depends partly on the nature of the information and partly on the preference of the author(s). However, it is crucial to present the various sections in logical order, so that a reader can follow the message without having to work too hard. Sometimes, arrows or pieces of text pointing out the next section are helpful to show the intended direction of flow of information. Although space is limited, you should cite references correctly and acknowledge any colleagues who helped you with either the scientific work or the compilation of the poster. Another common omission is the author's name and place of work or study! Since the point of the poster is to tell other people about your work and make contacts with people who may

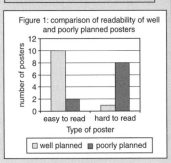

(a)

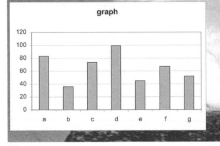

(b)

Figure 3.2 (a) A well-designed poster; (b) a poorly designed poster

be interested in helping you develop your ideas, it is crucial to make space for your name and contact details. It is worth bearing in mind that, as the author, you are trying to encourage someone else to take an interest in your work, so you must make it an appealing option for them to stop and read your poster. Figure 3.2a is an example of a well-planned poster, whereas Figure 3.2b shows a poorly designed poster.

3.5.4 Oral presentation

Giving a talk is a good way of reaching a wide audience. After a well-prepared and delivered talk, members of the audience should come away with a few memorable facts, which could help in their understanding of a topic or contribute to their own research. A really enthusiastic speaker can inspire their listeners to find out more about a subject which they previously had no knowledge of or interest in! At conferences there are usually a number of invited speakers who are nationally or internationally renowned experts in their fields, alongside talks from delegates at the conference who have applied for a short time slot (10 or 15 minutes) to give an oral presentation. This is often a useful way of presenting preliminary data or ideas which will be written up as a journal article later. At the end of each talk, there are opportunities for the audience to ask questions, which can open up interesting discussions. Although standing up in front of an audience is nerve wracking at first, most people get used to doing it. It is a good skill to have because through doing it, you become recognized and it is more likely that other people who are interested in your work will make contact with you.

Before you start planning your talk, confirm the remit that you have been given and the title you have agreed with the person who asked you to speak. Make sure you stick to these, as it can be very annoying for the organizer if a speaker talks about a different topic to the one arranged. At best, you might upset the audience who were looking forward to hearing about the original subject; at worst, you might be repeating information which another person will be talking about at the same event. During the planning and writing stages, it is therefore a good idea to check that your talk is not drifting off at a tangent from the planned title. If you think that you would be more comfortable talking about the new topic, this can be acceptable, but you must always discuss the change with the person who invited you to speak. Check the expected make-up of the audience, so that you use the correct level of scientific terminology and do not use abbreviations they might not understand. For example, an audience of consultant microbiologists would probably not appreciate it if you referred to vancomycin-resistant enterococci as 'small, round, drug-resistant bugs', but a patient group would probably prefer the latter phrase to the microbiologists' abbreviation of 'VREs'!

It is extremely important to keep to the allotted time when giving a talk. This is partly out of courtesy to the organizers of the event at which you are speaking. Speakers at conferences often fall into the trap of either packing too much into their talk or not including enough detail. Not only does it suggest poor preparation if a talk is much shorter or longer than the time allocated, but it might appear arrogant and unprofessional. Conference programmes tend to run on very tight schedules, which can be easily disrupted by speakers who run over time. If there are four speakers in the morning session and if every speaker takes an extra 5 minutes, by lunchtime your session will be 20 minutes behind! Similarly, if you are giving a short talk to colleagues in your department, they will have organized their working schedule to accommodate listening to your talk and may not be able to stay until the end if it takes longer than they expected. The Chair of an oral session will usually ask an over-running speaker to stop before they have completed their talk, which means the presentation is incomplete and can leave a poor impression on the audience. Another point to bear in mind is that by keeping to time, you are likely to be precise and concise in your explanations and discussions. If you are not succinct and instead start to waffle during your talk, the audience is likely to get bored and thus will not hear the message you are trying to convey. Make sure you present your information clearly and in a logical order during your talk. It is a good idea to start by reminding the audience of the title and your name and then provide an overview of the talk you are about to give. Organize your talk to include an introduction, the main part of the talk containing the key message(s), and then a conclusion or summary. If you are talking about a piece of research work, it is usually best to present your ideas in the same order that you would in a journal article – i.e. introduction, method, results, discussion, conclusion.

It may be useful to provide some background information for the audience to read while you are talking, e.g. on a PowerPoint presentation or overhead projector transparencies. This can help them to remember the key points you are making and also gives them something to focus on apart from you as the speaker. This may also help you if you are feeling nervous about being the centre of everyone's attention. However, remember that the idea of a talk is to engage the audience through words and pictures and your enthusiasm and interest for the subject. It is not necessary to write down everything you are going to say on your visual aids. The audience will then try to read all the written information on the slides, which will take their concentration away from what you are saying. If using PowerPoint slides, choose a simple design and if the background is patterned make sure that it contrasts well with the font and colour you have chosen for the writing. As for a poster, make sure that the background is not so bright or elaborate a pattern as to be a distraction for the audience. The font needs to be of a simple type face, e.g. Arial or Times New Roman, so that it is easy to read without straining the eyes. The type size for headings on slides should normally be 40–44 point and the main text between 28 and 32 point to be legible when projected on to a screen. PowerPoint slides should not contain too much information per slide. For written text, about three or four main points per slide are about right. If you are writing a slide and the programme changes the font size to be smaller than 28 point in order to fit in all your writing, that is a good indication that there is too much text on it. In such a case, you can either revise and shorten the points you are making or split your slide into two, clearer ones. It is always a good idea to summarize information or data in diagrams, tables or graphs, but make sure that they are labelled properly. Tables and figures should be numbered sequentially, i.e. Table 1, Table 2, . . ., Figure 1, Figure 2, . . ., and given clear, explanatory headings and suitably labelled columns and axes, respectively. Diagrams should have helpful annotations pointing out the salient features. It is usually better to put each diagram or table on a separate slide and it helps the audience to understand the points you are making with figures and tables if you interact with them during the talk. So, for example, you could walk over to the projection screen and point at an interesting piece of data with your arm or using a manual pointer. Alternatively, you could remain standing in the same place and use a laser pointer to highlight interesting parts of the table or diagram.

While giving the talk, try to appear confident (even if you aren't feeling so!) and look at the audience. If you feel nervous about making eye contact with people, pick a spot on the wall that you are facing, at the back of the room, and focus on that – the audience will think you are looking at them. Although it can be hard not to change your speech patterns when you are concentrating on them, you should not talk much faster or more slowly than normal and make sure you speak clearly. If you talk too softly, some of the audience may not hear you and, depending on the venue, you may have to project your voice to the back of the room – however, do not try to do this by simply shouting as loudly as you can, because your words still may not be clear. In many lecture halls, there is a microphone for the speaker to use. It is sometimes in a fixed position, in which case you need to stay close to it so that it will pick up your voice. In other places, the presenter is given a small portable microphone to clip to their lapel. This must not be too near your face or too far away – in both cases, your voice will be indistinct; it is best to get help with fixing it in a suitable position. If you have not memorized your talk or prefer to have the talk written out, it is better to use prompt cards, with the main points you want to remember, rather than reading directly from a prepared script if possible. Whatever you are reading, make sure you hold it at waist height at an angle of 45°, so that the audience can see your face and you are looking over it at them.

Whatever the subject of your talk and whoever the audience is, you should always be prepared to answer questions at the end. When giving an information talk to a users' group, there may well be points arising from your presentation which need clarifying, but bear in mind that other issues which are affecting users at that time might arise, since they have the opportunity to speak to someone from pathology. People are often very worried about the questions they might be asked during university presentations or at scientific conferences.

You can pre-empt this by deliberately skimming over a point during your talk and say ' I do not have time to go into detail about this, but am happy to take questions about it later', which directs the people in the audience to the question you would like to be asked! Whatever the nature of the question, if you can clarify the person's query then that can be very helpful, but if you don't know the answer, be honest and politely say so – you might feel awkward, but it creates a better impression than bluffing or waffling! Finally, remember that whether you have been asked to speak at a meeting or have submitted a short paper at a conference, which has been selected, you have been 'invited' to give your talk, so it is good manners and professional practice to thank your hosts for the opportunity. Box 3.5 gives a helpful check list of the points to remember when planning and giving an oral presentation.

Box 3.5   Giving an oral presentation

- Confirm the remit, title, likely audience and required length of the talk before you start planning.

- Present your information clearly and in a logical order, using clear headings for each section and sequential numbering for figures and tables.

- Use PowerPoint slides, overhead projector transparencies or hand-outs to convey the key points for the audience to retain.

- Choose the background, type face, font size and colour scheme of slides very carefully, so that they are a help rather than a distraction for the audience.

- Use only three or four main points per slide and put each diagram or table on a separate slide.

- Interact with your slides and the audience as much as you feel comfortable with; they want to know that you are interested in keeping their attention and are enthusiastic about the subject you are talking about.

- Speak clearly and look towards the audience when speaking; if you use prompt cards, do it discretely so the audience is not aware of them.

- Observe other people's talks, so you have a good idea of what is effective and what to avoid doing when you are giving your oral presentation.

3.5.5 Writing an article for submission to a peer-reviewed journal

When a scientist has done some work which has yielded interesting results, they usually want to share the information with a wide audience. Having a paper published in a respected and widely read journal is an excellent way of doing this. If you do write an article for a peer-reviewed journal, you are following in the best traditions of research scientists and biomedical scientists (see Chapter 1) in addition to undertaking a valuable continuing professional development activity (see above and Chapter 8). Scientists also spend a lot of their time writing applications for grants to allow them to conduct their experimental work and then compiling reports about the outcomes of those experiments. These usually have to be set out in a particular way, according the requirements of the grant-awarding body, and have similarities with the writing of a research paper. The reason for the standardized format is to make it easier for the people who have to read many applications and reports. Complying with the official constraints makes it more likely that the application will be given due consideration. Not all studies or projects are intended for publication and a local or preliminary study may be written up for distribution to a select group of people. Although not necessarily needing to follow any specific instructions for presentation, restricted communications of this type will usually follow the conventional structure of a scientific report, a familiar format allowing readers to concentrate on the content rather than being distracted by an unfamiliar order or layout.

The first step in preparing a journal article is to identify work which is of a high enough standard to be considered for publication. Research projects which form part of academic qualifications can meet the required level, if the work has been conducted with scientific rigour and the results are reliable and reproducible. Sometimes, a small 'in-house' study within the laboratory (for example, comparing a new test kit with the existing gold standard or changing laboratory practice) can yield results which are worth disseminating to other biomedical scientists. The decision about whether and what to publish should be taken in consultation with someone who has experience of writing and submitting academic articles. Similarly, it is important to select the journal to which you will submit your article carefully; a balance has to be struck between the prestige of the journal, the wideness of its readership and the likelihood that it will consider a paper on the subject you have researched or reviewed. Therefore, you should always ask the advice of a university tutor or senior laboratory colleague before proceeding, and usually projects are carried out in conjunction with an academic or clinical laboratory supervisor or as part of a team of researchers who will be involved in discussions about these aspects.

Each journal has its own writing and presentation style, so once you have chosen a particular journal, it is a good idea to read some of the articles which they have already published, to familiarize yourself with the required format before starting to write yours. Although there is some variation between the requirements of the various publications, the basic structure includes a short summary of the article and sections detailing the study background, methods used, results, discussion of the findings and a list of references (see Box 3.6). The journal's required format for submission of manuscripts will be detailed in a set of 'authors' instructions', which can usually be found in a full copy of one issue of the journal or via the journal website. These will specify details such as word limits, style of headings, page layout, margins, font and font size; if you establish these for yourself at the beginning of the writing process, it saves time and frustration at the end. A very important point to check is the way in which the journal asks for references to be cited. In the text, it is usually the Vancouver style of superscript numbers (e.g. Research shows that cats prefer to sit on mats[1]) or the Harvard style with authors and years [e.g. Research shows that cats prefer to sit on mats (Smith and Jones, 2005)]. There is often a specific format for listing references at the end and this must be verified. For example, some journals ask you to give the names of the author(s) in capital letters, others require the volume of the journal to be given in bold type. In some cases, the style is for the whole journal name to be written in full and in italics, whereas in others, they want you to use accepted abbreviations, such as *Br J Biomed Sci*, standing for the *British Journal of Biomedical Science*. The accepted list of abbreviations is publicly available, for example on the Web of Science database. It is always the submitting author's responsibility to check the authenticity and accuracy of the references and to present them in the required format. A well-written account of an excellent piece of scientific work could be rejected by a journal if the author has not paid attention to detail in their presentation of the references.

Box 3.6 Writing an article for submission to a peer-reviewed journal

- Select the material to write up and a suitable journal carefully, in consultation with the academic or clinical laboratory supervisor or research team colleagues.

- Familiarize yourself with the journal's style, their formatting arrangement and required method for citing references.

- Plan the article so that it flows in a clear and logical order. Start with an overview and working title and refer back to these as you are writing; if you are straying from the original plan and title, revise them or your text as necessary.

- Write succinctly, using perfect grammar and spelling, defining abbreviations and avoiding jargon wherever possible.

- Use diagrams, tables and figures to summarize information or data as appropriate; make sure that they are numbered sequentially and have clear headings and labels.

- Ask someone whose judgement you trust to read your first draft and take their advice seriously. You will probably find that you further revise the article several times before finally submitting it to the journal's Editor for consideration.

- Check the final version of your paper thoroughly for scientific accuracy and to ensure that you have followed the required journal guidelines before sending it in. Make it easy for the Editor to process your paper through that journal's peer review system, so that your work receives the consideration it deserves.

- Be prepared to wait patiently before finding out whether the journal has decided to accept your work for publication and remember that you are not allowed to submit it to another journal in the meantime. However, if you have not heard anything from the journal for several months longer than the stated time for their peer review process to be completed, do not hesitate to contact the Editor.

Once you have established the formatting style, you can then start on writing the article itself. You should begin by thinking of a suitable title which describes your work and the main messages which you want to convey in the paper. Write out everything you want to say in the paper in a logical order. Plan each section so that the information flows in a coherent order and is easy for the reader to follow. As you are writing, keep checking back to the title and the section plan at regular intervals to make sure you have not drifted too far from the original idea. If you have, it may be that you need to revise what have been writing, to make sure that you keep to your plan and avoid confusing potential readers. However, it is not unusual to find that as you have started to clarify your thoughts, your original ideas or interpretation of your results have changed and it is the title and the angle of the article which need to be altered. This is acceptable, provided that you keep checking for consistency within the article.

In addition to the journal's prescribed format, you should pay attention to your own writing style. Always try to write clearly and succinctly and do not use big words and long sentences just to show how clever you are – it rarely has the desired effect! In order to avoid any confusion, it is good scientific and professional practice to define abbreviations and technical terms the first time they are used in the article. It is also a good idea not to use jargon words at all, but if you think you need them to help convey your message, then make sure that you explain them carefully, in a way that a non-specialist reader could understand. It is important to ensure that your grammar and spelling are of the highest possible standard, to prevent any misunderstanding. Consider, for example, the following: 'The patients results suggested liver disease – e.g. there LFT's where abnormal'. There are a number of simple and common errors of grammar, punctuation and spelling in this, which combine to make a very confusing sentence. Care and proof reading could have eliminated the mistakes, but as it stands, it is not clear how many patients are involved or the location of the abnormal liver function tests! Compare this with the correct version: 'The patients' results suggested liver disease – i.e. their LFTs were abnormal', and notice how small errors could lead to large misunderstandings. It is always a good idea to summarize information or data in diagrams, tables or graphs, but they are only useful if they are labelled properly. As already mentioned, tables and figures should be numbered sequentially – Table 1, Table 2, . . ., Figure 1, Figure 2, . . ., and given clear, explanatory headings and suitably labelled columns and axes, respectively. Diagrams should have helpful annotations pointing out the salient features. It should be possible for the reader to understand the key points of information that you are presenting in each table or diagram without reading the rest of the article. Within the text, you should explain the significance of and interpret the data you have presented in any figures or tables.

Once you have started writing, ask someone whose judgement you trust whether they would have time to read your first draft and help you improve the work. Give them plenty of notice and agree a reasonable time scale for them to provide their feedback. It could be helpful to set a meeting a few weeks after you have sent them the draft (but remember that in the meantime, you should be patient and recognize that people do not generally respond well to constant enquires as to whether they have read your work!). Take the advice that this person gives you seriously and work with them on subsequent drafts. You will probably find that you have to revise the article several times before finally submitting it to the journal's Editor for consideration. When you are finally happy with the article and ready to send it in to the journal, check again that you have followed all the guidelines. It is common to be asked for tables and figures to be compiled separately and sometimes the text has to be double spaced. Verify the correct format for submission; for example, some journals are happy to work with an electronic copy sent by e-mail to the Editor, but some still ask for a set number of printed copies or an electronic copy compatible with a particular program. All these formatting 'rules' are in place because, after the Editor has decided whether the subject matter is suitable for the journal, it will be sent to anonymous reviewers for consideration. The format and spacing of the text are designed to make it easier for the reviewers to read the work and add comments. Your paper might arrive on the Editor's desk or in their e-mail inbox on the same day as 10 other articles, not all of which could possibly be accepted. If you have followed the guidelines meticulously in your article, it is much more likely that your work will be given due consideration.

It is worth finding out from the Editor how long the journal usually takes before it communicates to you whether it will publish your paper or not. It is often several months, so patience is needed. However, if you have not heard for a long time, it is a good idea to contact them to see what progress your article is making through their peer review system. Remember that while your paper is under consideration with one journal, you cannot submit it to another. A proportion of articles are eventually rejected for a number of reasons, but they are usually sent back with the reviewers' comments which help you to amend or re-write the paper. Once it has been sent back to you, it is your decision how to proceed – sometimes the journal invites you to re-submit a revised version, but you can equally send the article to a different journal for consideration and start the whole process again. If you do that, make sure you format it according to the new journal's requirements. Once your paper has been formally accepted, it may still be several months before it appears in print, but the journal has agreed to publish it in a future volume when space is available, so you can feel suitably proud of the achievement at that point!

3.6 Conclusion

The ability to communicate the right information, clearly and succinctly, to the person who needs to know it, in good time, is an important skill for a biomedical scientist to develop. As this chapter has shown, effective communication requires careful planning and plenty of practice. Consider the tips and examples in the Boxes carefully and have them in mind next time you look at someone else's poster or listen to a professional talk. You can learn by appraising other people's work – make sure you follow good practice and do not repeat mistakes!

Quick quiz

1. Name three professional groups with which biomedical scientists communicate regularly.

2. List three of the principles of effective communication.

3. Is an e-mail usually a formal or informal method of communication?

4. Imagine you have some preliminary data from a study comparing a new diagnostic test method with an existing and widely used one. To whom would you communicate your findings and by what means?

5. Why is it important to follow scientific conventions such as SI units and professional protocols such as SOPs during communication?

6. What is the role of peer review in science?

7. How could a newly graduated biomedical scientist disseminate the results and conclusions from their undergraduate research project?

8. How is standard 7 in the Health Professions Council Standards of Conduct, Performance and Ethics relevant when communicating as a biomedical scientist?

9. What would be the most effective way to inform users of a department's service that the laboratory will be staying open an hour later in the evenings for collection of routine samples?

10. What are three elements of a well-planned poster?

11. Why should you never exceed the allotted time for an oral presentation, particularly at a conference?

12. What is the correct way to cite references in a journal article?

Suggested exercises

1. Think about the information which staff on the wards and in primary care who are new users of the service provided by your Pathology Department might need. Prepare a 10 minute talk for a multiprofessional audience, which includes newly qualified practitioners (e.g. junior doctors) and staff who have recently joined your Trust from another one (e.g. a midwife who used to practice in another part of the country).

2. Imagine that your laboratory is about to introduce a new test method for an analyte or antibody. This new method requires the use of serum, rather than plasma, as used for the test it is intended to replace. This will necessitate a change in the blood collection tubes required for specimens sent to the laboratory for this particular test. Plan your communications with the relevant staff in pathology and service users and decide how best to disseminate the information, so that everyone who needs to know about the change receives the message.

3. Following the guidelines within your laboratory, prepare a written SOP for a newly introduced test method. Check its effectiveness by asking

a colleague who is unfamiliar with the method to carry out the test using only your SOP. Repeat this exercise by preparing a SOP for the maintenance of a newly installed piece of equipment.

4. Choose a scientific article that interests you from a peer-reviewed journal. Prepare and give a presentation for your department's journal club.

Suggested references

Clayton, J (2000). Written Communication Skills. *Biomedical Scientist* **44**.

Lo, Y. M. D., Chiu, R. W. K., and Chan, K. C. A. (eds) (2006). *Clinical Applications of PCR (Methods in Molecular Biology)*, 2nd edn. Totowa, NJ: Humana Press.

Matthews, J. R., and Matthews, R. W. (2007). *Successful Scientific Writing*, 3rd edn. Cambridge: Cambridge University Press.

Saiki, R. K., Gelfand, D. H., Stoffel, S., Scharf, S. J., Higuchi, R., Horn, G. T., Mullis, K. B., and Erlich, H. A. (1988). Primer – directed enzymatic amplification of DNA with a thermostable DNA-polymerase. *Science*, **239**, 487–491.

Theaker, J. (2007). Polymerase chain reaction. *Biomedical Scientist*, **51**, 531–533.

www.ibms.org/Home/Science/Articles/BiomedicalScientistArchive.

4

Quality management in the clinical laboratory

4.1 Introduction

Pathology test results can play a vital role in the diagnosis, monitoring and treatment of patients, provided that they are as accurate as they possibly can be. This means that biomedical scientists and their colleagues in the laboratory must work to the highest possible scientific, technical and professional standards at all times. The correct specimen type must be collected in the appropriate way from the right patient; it must be processed accurately and tested using the expected assay; and the result must be recorded precisely and reported to the correct user in time to be useful for the patient's care and treatment. Incorrect procedures at any stage in the passage of a specimen through the clinical pathology laboratory can lead to serious consequences for patient safety and care, as illustrated by some examples in Table 4.1.

To ensure that the highest possible standards are met in all areas of work in a clinical laboratory, biomedical scientists and their colleagues continually monitor their practices and procedures. This chapter describes the key elements of the systems used, together with a consideration of the factors that affect the quality of work in the pathology laboratory.

4.2 Quality in pathology

To ensure that the highest possible standards are met in all areas of work in a clinical laboratory, biomedical scientists and their colleagues continually monitor their practices and procedures. This includes checking the day-to-day running of test reagents and equipment within the laboratory, and also evaluating standards against national (and sometimes international) external measures. In a modern diagnostic pathology department, making sure that

An Introduction to Biomedical Science in Professional and Clinical Practice   Sarah J. Pitt and James M. Cunningham
© 2009 John Wiley & Sons, Ltd

Table 4.1 Examples to illustrate the importance of following correct procedures at each stage of specimen passage through a clinical pathology laboratory

Specimen passage stage	Example	Special requirements for procedure	Possible errors
Specify sample type	Plasma sample required	Collect blood in appropriate anticoagulant blood tube	Anticoagulant not used; blood clots and test procedure cannot be followed
Collect sample	Fasting blood glucose	Collect sample in early morning	Sample collected shortly after patient meal; incorrect result
Record patient details	All requests	Provide minimum data set (see Chapter 5)	Insufficient patient detail recorded; patient cannot be identified
Process sample	Viral load test	Process blood sample and freeze plasma immediately to minimize analyte degradation	Storage of sample at room temperature; incorrect result
Analysis/ examination of sample	Morphological examination of kidney biopsy	Use appropriate histological stain, e.g. haematoxylin – eosin	Incorrect stain used, e.g. Gram stain; appropriate examination cannot be carried out
Record test result	Fasting blood glucose	Quantitative result in mmol/L	Incorrect form of result reported, e.g. 'positive' or 'negative'; result cannot be interpreted
Report result to user	Follow-up blood sample analysis for patient admitted with chest pains	Clear report format with additional comments as appropriate; report immediately to patient's ward	Result returned to incorrect location, e.g. Accident and Emergency where patient admitted and relevant clinical comments omitted; delay and inappropriate treatment

all the required standards are being met and that all the quality systems are working effectively is a complicated task. Many laboratories now designate one of their senior specialist practitioner biomedical scientists as a 'Quality Officer', although sometimes the Laboratory Manager takes on or shares this role. It is also increasingly common to have an advanced practitioner (Band 8) as overall 'Quality Manager' for pathology, to coordinate and support each individual department's work. However the work is divided, it is important to have a 'Quality Management System' (QMS), which has been designed by a representative group of staff from the individual laboratory or overall pathology department; this is to ensure that the system is feasible in that particular working environment, according to the nature of the discipline and the mixture of staff. Box 4.1 lists the key elements to maintaining quality that biomedical scientists must be aware of, and each aspect is discussed in detail below.

Box 4.1   Key definitions in pathology quality management

- **Quality control**
  Checking the performance of a particular test each time it is run; this is to ensure that the operator, reagents and equipment are all working properly.

- **Quality assurance**
  Overall term for monitoring all practices and procedures within the laboratory to make sure that required standards are being maintained.

- **Quality assessment (internal or external)**
  Testing the standard of the laboratory procedures usually by testing a 'blind' sample. This can be an internal sample or one supplied from an external laboratory (in this case, the department can measure its performance against national standards).

- **Quality audit**
  Internal check of the whole procedure from the patient, through the laboratory and back to the patient; exercises can involve members of staff outside of pathology.

- **Clinical governance**
  A system of regulation and monitoring to make sure that the highest quality of patient care is provided within a healthcare organization.

- **Quality management system**
  Overall system of procedures and policies intended to ensure that defined standards are met, with clear plans to address failures to meet these standards (the documentation in this system must be 'controlled').

- **Accreditation**
  Procedure through which a department is given formal recognition that they meet certain minimum standards in their work, as defined by an independent organization.

4.3 Quality Control

When conducting any scientific experiment, it is important to run 'controls', which are samples containing known amounts of the analyte being tested for. If there are any problems with the test, such as an error in the dilution of a particular reagent, then the controls will not give results within the expected range. This will alert the scientist to the occurrence of a fault, although not necessarily directly to the actual cause, which may need some careful detective work! This system of using controls is equally important in routine diagnostic laboratory tests to ensure that any result is credible and accurate. In some diagnostic assays, a series of controls with staged amounts of the analyte under investigation ('standards') are put through the test and the results are used to construct a calibration curve, from which readings from samples with an unknown concentration of the substance in question can be compared.

These processes all form part of Quality Control (QC), which is important in the work of any laboratory, but it is especially significant in clinical laboratories, where the results relate to patients. QC in clinical laboratories was first introduced in disciplines such as clinical chemistry in the 1950s, where the assays that laboratories were using required controls for calibration. The statistical methods developed in industry by Shewhart, to monitor the performance of assays and equipment, were adapted for use in clinical chemistry laboratories by Levey and Jennings in the 1950s and are still used in some QC procedures today (see Figure 4.3). During the 1950s and 1960s, laboratories of all pathology disciplines began to monitor the performance of their equipment and reagents, in ways that are now accepted as routine. For example, each time a new batch of a particular bacteriological culture medium is made, it is tested with a control organism known to grow on it, to ensure that the medium had been prepared correctly. Similarly, to check that a stain is of the correct chemical composition, dilution and pH, laboratory staff test it on slides of tissue from organs with a known pathology or blood showing particular haemoglobinopathies and white cell abnormalities. When serological tests do not require full calibration, samples which are known to be positive and negative in that particular assay are included in the run. All these checks would be regarded as 'good practice' or indeed simply 'common sense' to a well-trained scientist, as it is not always possible to detect technical errors. Commercially available serology test kits include positive, negative and sometimes borderline control samples which are treated in the assay in the same way as patients' specimens. Similarly, slides containing cells of known morphology or pathology, are provided with staining kits, to check the performance of the stain. Many laboratories also include their own 'in-house' controls regularly. An assay or stain which is not functioning

at its optimal level can still produce a result, but its sensitivity or specificity (see Section 6.7, Evaluation of a diagnostic test, in Chapter 6) could be
compromised.

Selected examples of the Internal Quality Control (IQC) procedures used for
different aspects of clinical laboratory work are given in Boxes 4.2–4.4 and
Figures 4.1–4.3.

**Box 4.2   Internal quality control in antibiotic sensitivity
testing**

When a pathogenic bacterium has been isolated from a patient's specimen, it must be tested against particular antibiotics to decide whether it
is sensitive to them (thus making them suitable for treating the patient's
infection). In diagnostic laboratories, this is usually carried out using
paper discs which have been impregnated with a known amount of
each antibiotic under investigation. A culture of the bacterial isolate is
spread evenly across an agar plate and discs are placed on top of this.
The antibiotic diffuses into the agar, so that if that bacterium is sensitive to that antibiotic, then it will be unable to grow across the plate;
where the concentration of antibiotic is sufficient to prevent the growth
of the bacteria, then areas where there is no growth ('zones of inhibition') will be observed. The diameter of these zones can be measured
(for example, with a small ruler) and compared with results obtained
from a control organism which is known to be sensitive or resistant to
that antibiotic. There are two main methods for making the decision
about whether an organism is sensitive to a particular antibiotic. In the
Stokes' method, the test and control organism are grown together on
the same plate, so that any variations in culture conditions will affect
both equally (see Figure 4.1a). The diameter of the zones of inhibition for each antibiotic are measured and compared; thus the internal
quality control is achieved by treating the test and control organism in
the same way at the same time. The Kirby Bauer method uses a separate agar plate for each test isolate and each control organism; they
are cultured under carefully controlled conditions and the diameter measurements are compared with results from a known control organism.
Thus a set of tests could involve the use of one control plate to be
used with several plates from organisms with unknown sensitivities (see
Figure 4.1b).

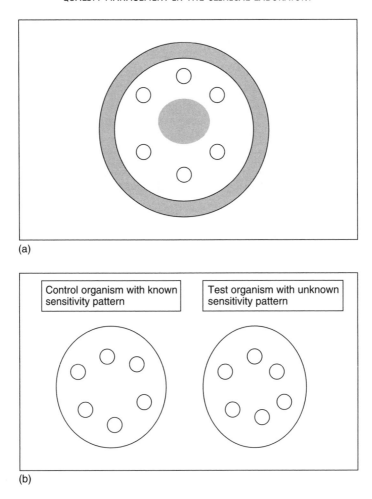

(a)

(b)

Figure 4.1 (a) Diagram to show arrangement of plate in Stokes' method for antibiotic sensitivity testing. Shaded areas indicate inocula-test (inside) and control (outside). The quality of the medium and growth conditions are controlled for by comparing the sensitivity pattern of the control organism directly with that of the test; any variations in zone size caused by the culture environment will affect both organisms equally. (b) Diagram to show arrangement of plates in the Kirby Bauer method for antibiotic sensitivity testing. If the control organism shows expected pattern of sensitivity to the antibiotics, then the quality of the medium and the growth conditions are taken as satisfactory for all test plates and therefore that the results for the test organisms are reliable

Box 4.3 Internal quality control of stained slides

It is good practice to use a control slide containing cells of known morphology or pathology when undertaking staining of slides. Controls should be included regularly (e.g. daily or weekly depending on how often the stain is used) and every time a new batch of stain is prepared. In this way, the quality of the stain can be determined. If the stain has been made up to be too strong, too dilute or at the wrong pH, then the appearance of the cells on the control slide will highlight this; similarly, if the stain was prepared some days previously and has deteriorated, then the characteristics of the control cells will not be as expected and readings made from test slides will not be reliable. Figure 4.2 shows some thin blood films which were stained with Giemsa stain; note how the variability in the quality of the preparative agents and the stain can affect the observer's ability to identify cells and detect abnormalities.

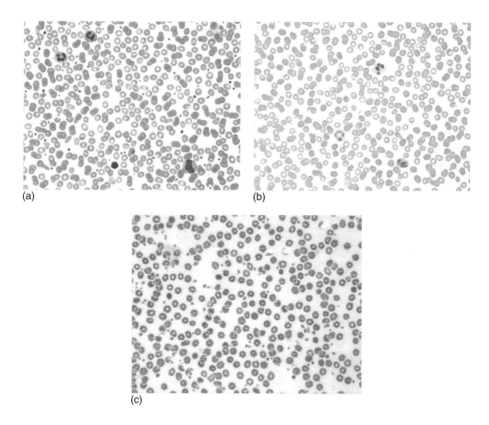

(a)

(b)

(c)

Figure 4.2 A series of thin blood film slides: (a) prepared and stained correctly, allowing appearance of red and white blood cells to be clearly seen; (b) stain prepared at too low pH, so cells appear too pale and cellular structure is understained; (c) methanol used to fix slide was contaminated with water so red cell morphology is distorted. Photographs courtesy of Mrs Sarah Bastow

Box 4.4   Internal quality control in serology

Serological assays, whether manual or automated, usually require the inclusion of positive, negative and calibrator samples in each run (or regularly). In commercial kits, these controls are provided by the manufacturer and are designed to meet certain specifications. They are used to calculate whether certain parameters have been met during the assay run (for example, that the negative sample gives a reading below a certain value) and, if they have not, then the run would be rejected. However, it could be possible for an assay run to be accepted by the kit's internal calibration and still give inaccurate readings. In the example of the requirement for the negative sample to provide a reading below a certain value, if there was a problem with the assay which caused all the readings to be too high, then the negative control would be too high and the assay would be rejected. If, however, the fault had caused the readings to be too low, this might not be shown up through the performance of the negative controls; test samples which did contain that analyte, but at a low concentration might be falsely recorded as 'negative'.

In order to monitor the performance of an assay over a period of time, it is usual practice to include 'in-house' controls, which are made up from pooled patient sera. The mean value for the analyte and the standard deviation (SD) from the mean are calculated, by running an aliquot of the pooled sera in the particular assay a number of times (e.g. 20 times). These data are used to construct a Levey – Jennings (or Shewhart) plot on which ± 1 SD, ± 2 SDs and ± 3 SDs are marked, as shown in Figure 4.3.

This 'in-house' sample is then included as a test sample each time the assay is run. The result obtained is recorded on the graph and this allows easy comparison with values noted from previous runs. This visual representation will show that there might have been a problem with that run of the assay, even if the kit's internal controls give values within expected ranges. In this case, the biomedical scientist has to consider how to proceed. For example, if the 'in-house' control is slightly out of range, they might accept the results for today's assay, but keep an eye on future performance; if the 'in-house' control gives a value which is considerably different from expected, they might have to reject today's assay, make up a new batch of reagents and test all the samples again.

There are two types of error which can occur, namely 'systematic' and 'random' error. Systematic error is the description of a situation where the values for the 'in-house' control move gradually out of an acceptable range and this usually takes place over several runs. On the Levey – Jennings plot, this kind of problem is noticeable because the values recorded will all fall on the same side of the mean. This should not happen since the mean should be 'in the middle', implying that some days the 'in-house' control will be above the mean and some days it will be below. In Figure 4.3, the results from days 4–7 are indicative of possible systematic error. When this kind of error occurs, the plot also often shows that the result for the control is increasingly greater or less than the mean each time, i.e. the results for the sample are drifting away from the mean. This commonly happens because the reagents are losing their potency, but sometimes the results for the kit control have drifted in the same direction and the internal calibration may not fail. In this case, the 'in-house' control, which has a known expected value, is very useful to draw attention to the problem. Random errors, by contrast, usually provide isolated results which are out of the acceptable range and often (but not always) very obviously different from previously recorded values. The result for day 11 in Figure 4.3 is clearly a random error. These usually indicate that the person doing the test has made a mistake in the procedure or that the equipment has a malfunction.

The biomedical scientist can make decisions about whether to accept or reject the run for that assay on that day by applying a series of rules called 'Westgard rules'(www.westgard.com). The rules make use of the SD values marked on the Levey – Jennings plot (see Figure 4.3) and are summarized in Table 4.2. These take the biomedical scientist through a series of steps before concluding whether to accept or reject that particular run. If the decision is made to repeat the assay, looking at which Westgard rules have been broken can help to point to the nature of the technical problem (for example, are the reagents going off or has the machine broken?) The plot is an important tool for the Quality Manager to monitor the performance of that assay over a period of time as part of Quality Assurance (see Section 4.4).

4.4 Quality Assurance

While QC concentrates on the scientific and technical performance of an individual assay, there is more to providing a high-quality diagnostic service than accurate results. The correct specimen must have been prepared properly and put through the right assay, and the result must have been reported in the best format to the appropriate person in good time. Looking at the practices and procedures within the laboratory to ensure that a good service is being provided is called Quality Assurance (QA). Monitoring all areas of the work in the laboratory is an important part of QA and it is helpful to divide up the processes within

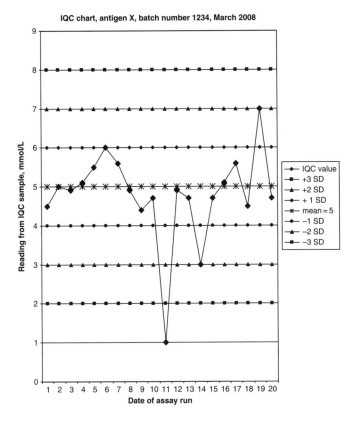

Figure 4.3   Example of a Levey – Jennings plot showing daily recordings of values obtained for an IQC sample in a serological assay, where mean = 5 mmol/L, SD = 1. Note how variations in assay performance during the month of March are clearly shown by this visual representation

the laboratory into three categories or phases – 'pre-analytical', 'analytical' and 'post-analytical' (these are described in detail in Chapter 5).

The 'pre-analytical phase' comprises all the steps from when the test request is generated from a particular patient, through to the sample being taken to the bench within the laboratory to be tested. The first parts of this process take place outside the Pathology department; staff in pathology take it on trust that their colleagues on the wards and in the general practice surgeries have collected the right specimen from the correct patient and have explained to them the reason for the test request. The most suitable specimen container must be used, which must be sealed and then placed in a plastic bag along with the request form. Once collected, the samples should be taken to the laboratory in good time (or stored appropriately until they can be transported), but also with care so that containers do not break. It is clear that although all these parts of the pre-analytical phase have an impact on the quality of the laboratory work, pathology staff have little influence over them on a daily basis. Therefore, as

Table 4.2 Westgard rules

Rule	Shorthand	Description	Type of error indicated
A	1 $_{2\,SD}$ rule	If the control value is more than ±2 SD from the mean, then check the previous value and consult rules B–F. If the control value on the previous run was satisfactory and other rules are not broken, then accept the run	
B	2 $_{2\,SD}$ rule	If the control value is more than ±2 SD from the mean (*on the same side* of the mean) on two consecutive runs, then it should be investigated before deciding whether to accept the run or not	Possible systematic
C	4 $_{1\,SD}$ rule	If the control value on four consecutive runs is more than ±1 SD from the mean (*on the same side* of the mean), then this suggests that a recalibration or service of the equipment is required; the run still might be acceptable depending on the circumstances	Systematic
D	10× rule	When 10 consecutive control values are on the same side of the mean (regardless of whether they are within an acceptable range), then the run should be rejected. This indicates a drift, which is often due to a decline in the efficacy of the reagents	Serious systematic
E	1 $_{3\,SD}$ rule	If the control value is ±3 SD from the mean, then the run should be rejected. This is usually caused by a random error such as a mistake by the operator, wrong concentration of a reagent or problem with the incubator; however, it could also be part of a systematic trend	Systematic or random
F	R $_{4\,SD}$ rule	If the difference between two consecutive control values is more than 4 SDs, then the run should be rejected, as this indicates a serious random error	Serious random

Adapted from www.westgard.com.

part of QA, the Quality and Laboratory Managers have to work closely with their colleagues who have direct contact with patients, to make sure that there are clear written instructions about how to collect and transport specimens. All staff involved need to be trained properly and must understand why each step is important. Once specimens have arrived in the laboratory, it is easier to monitor what happens to them and to ensure that all staff are trained and equipped to work appropriately and safely. For example, all staff should have access to personal protective equipment and there should be clear written procedures on how to deal with leaking or broken sample containers. Allocation of laboratory numbers can be complex, especially if some types of specimen use a different number series from others, so the system must be clear (e.g. colour coding, using variations in font or label size). Receiving samples and preparing them for the processing in the laboratory are a key part of the pathology service. If mistakes occur at this stage, the wrong test could be conducted on the wrong sample from the wrong patient! Therefore, it is important that specimen reception staff are supported to do their job well and that their role is valued by all biomedical scientists.

The 'analytical phase' involves the scientific and technical aspects of the laboratory's work. The QC techniques outlined above are a core element to QA of the analytical phase; the Section Leader and Quality Manager should assess the results regularly, regardless of whether any problems have been reported. They must also be involved in 'troubleshooting' when any of the QC criteria have failed in a particular testing method. All biomedical scientists can participate in QA, by being careful to follow standard operating procedures (SOPs), not carrying out any tests or using any equipment which they have not been adequately trained to use and reporting any mistakes they make, immediately, to a senior colleague. Although it seems like a boring chore, regular checking and recording of the temperature of refrigerators, freezers, incubators and water baths is a very important part of QA, since for some reagents, a 1 °C difference in storage temperature could detrimentally change their activity. Regular review of the testing methods used, to make sure that the procedures are relevant and up to date, is a necessary part of monitoring quality. It is also essential to have a system for all staff to record incidents or accidents and log errors. The Quality Manager can look at these to see if there are any patterns which could indicate that a review of health and safety procedures is required or better training of staff in a particular aspect of the laboratory work is needed. The quality of the results sent out can be assured by only employing staff with the correct qualifications and experience in each role and by supporting them to do the job well. This includes providing full training for each task, personal protective equipment and ensuring that they are aware of all relevant health and safety issues. Staff are also more likely to work well in the laboratory if they have job satisfaction and are able to develop their careers, for example by having the opportunity to take on new roles and responsibilities or take an educational qualification. This should be carefully planned and all training and education activities undertaken by each member of staff should be recorded.

In terms of the quality of the service to users, the 'post-analytical phase' is the part which they are most likely to notice, so it is important to take care over it. The way in which urgent results are telephoned through to the ward or general practice surgery must be considered, since it is important that the correct information is conveyed to the doctor or nurse treating the patient; by having a polite and clear telephone manner, biomedical scientists can show that staff in pathology are professionals in whom the clinicians can have confidence (see Chapter 3). The written version of the result must contain all the relevant information which might be needed (i.e. clear patient identification and clinical interpretation of result, if necessary, along with the laboratory findings). Part of an assessment of the quality of the report would be how easy it is for the person reading it to find the information they need while the patient is with them. This does need to be checked, as sometimes the format on the screen is hard to read and even a well-designed page on the computer does not always translate into a clear printed version, as the examples in Figure 4.4 illustrate.

In order that pathology tests contribute to patient care, not only must an accurate result be sent, for the right person, in a legible format, but it must also arrive in time to be useful to the clinician making decisions about the diagnosis and

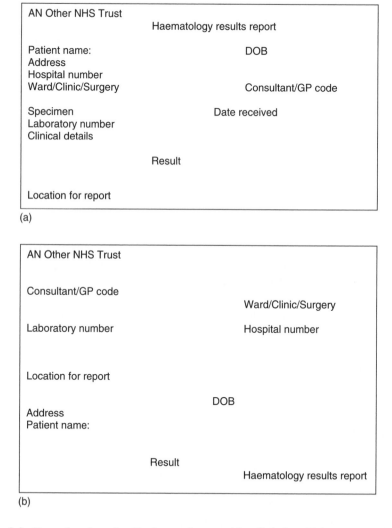

(a)

(b)

Figure 4.4 Examples of good and bad report formats: (a) well-designed laboratory result report, where information is presented in a logical order and the result is prominent; (b) poorly designed laboratory result report, where information is presented in a haphazard order and the result is hard for the reader to find

treatment. The length of time taken from the specimen arriving in the laboratory to the result being available is called the 'turnaround time'. Turnaround
times are often agreed by Pathology Managers in their discussions with users
and can depend on the nature of the pathology investigation and the urgency
of the result. Whereas routine histopathology examination and bacterial culture
take several days and cannot be done more quickly, serology tests may take
less than an hour when performed on an analyser. It would be a good use of
resources to validate and send out a urea and electrolyte result within a hour
for a patient in Accident and Emergency. However, this might not necessary for
a routine antenatal screen for Rubella antibodies; it could be agreed that such
specimens would be tested twice weekly in batches, giving a turnaround time of
a week. Canvassing the opinions of the users of the laboratory on such matters
as report format and turnaround times are a key part of monitoring the quality
of the pathology service.

Although QA is a particular remit of the Quality Manager, all biomedical scientists can contribute to assuring the quality of the service delivered by their
laboratory by working conscientiously themselves, helping their colleagues and
participating in audits (see Section 4.6). When a problem with quality is identified, it is important that all members of staff are consulted in the efforts to
remedy it. They need to understand why the error occurred, the consequences
of the error and how to prevent it happening again. Since working in a Pathology
Department is a 'team game', it should be obvious that a mistake is rarely due
to the actions of one person. The person who performs the wrong test may
have done this because they have not been supervised properly or have been
asked to do the assay without the correct training. This often happens because
the department is short staffed or the supervisor has felt pressured into dealing
with another issue. The idea behind QA is to maintain and improve standards;
this can only be done with the cooperation of all staff. An individual should
not be singled out as the scapegoat for a particular incident or mistake, but
everyone should be encouraged to help and support each other to do their best
(see Section 4.10).

4.5 Quality assessment

4.5.1 Internal Quality Assessment (IQA)

It is usual to have a system of Internal Quality Assessment (IQA) within
diagnostic laboratories. This generally entails taking an aliquot of a specimen
from which the result has already been reported and re-introducing it into the
laboratory as if it were a new sample. One person, usually the Quality Manager,
removes the previous laboratory number and creates a new request form, so
the sample is re-tested 'blindly'. The idea is to ascertain whether the same

result is obtained and the report sent out within the same amount of time as previously; thus it is a check on the whole process from specimen reception to reporting. The findings of each IQA exercise should be discussed with the staff involved (e.g. during a staff meeting). If a discrepancy is found between the report originally sent out and that generated from the blind re-running of the sample, then reasons should be explored, within a 'no-blame' environment (see Section 4.10). This exercise is useful because it can highlight problems which have not been noticed or considered a high priority. For example, an inconsistency might be noted between results for a test which is performed once per week and requires a reagent which must be kept at $+4\,°C$ after the bottle is opened. If the first test gives a higher reading than the second, then the integrity of the reagent may be suspected. An investigation of the temperature in the refrigerator might find that the thermometer is broken and that although daily temperature readings have been recorded, they are inaccurate, meaning that the reagent has been stored at about $+10\,°C$ (instead of $+4\,°C$) between the first and second tests and had therefore deteriorated. Another example might be a difference in the time taken to produce printed copies of the results. If the first result was ready in 72 hours, but the second result was printed within 24 hours, that would be an indication that there was a backlog of reports waiting to be printed at the time of the validation of the first result which was resolved before the blind re-test. Problems such as these might not be highlighted during a formal audit during which all the documentation is inspected; in the example of the problem with the refrigerator temperature, the daily thermometer readings were recorded, so the part of the audit looking at the equipment monitoring process would have been fulfilled.

4.5.2 External quality assessment (EQA)

Schemes for challenging a diagnostic laboratory's processes by testing a specimen from an external source were first introduced during the late 1960s and early 1970s in the United Kingdom. The idea was to make sure that all the pathology laboratories in the country could accurately measure analytes, correctly identify microorganisms or recognize abnormal cells in blood and tissue samples, as applicable to each discipline. The National External Quality Assessment Schemes (NEQAS) for haematology, clinical chemistry, histopathology and microbiology were each originally set up to help laboratories provide results that were of good enough quality to be clinically useful. The idea was to assist each laboratory in the country to reach and maintain a minimum standard of work, rather than vilify departments which did not. Although the original schemes have become more taxing over the last 40 years and have been joined by new schemes, reflecting the increasing range and complexity of pathology tests, this ethos of supporting rather than denigrating still exists.

There are now over 140 NEQAS schemes, each sending out specimens for testing for an individual analyte (e.g. red cell volume) or related groups (e.g. cardiac markers). They are arranged into seven discipline groups, namely andrology, chemistry, genetics, haematology, histopathology, immunology and microbiology, although the schemes within each group may be run by different centres (www.ukneqas.org.uk). For example, the microbiology group comprises the bacteriology, virology, parasitology, mycology and antibiotic assay schemes; the first two of these are organized by one centre, whereas samples for the last three come from separate specialist centres in different parts of the UK. Individual laboratories can sign up to receive samples from as many or as few NEQAS schemes as they require, depending on the repertoire of tests they perform. There is an annual fee to join each one and, in return for the money, the laboratory receives sets of 'unknown' specimens regularly (usually 3–5 times per year) plus information and support on quality issues. The 'unknown' specimens are prepared from patients' sera and tissues or microbial cultures at the NEQAS centre in such a way that they will provide an intended result when processed correctly. Thus, the NEQAS centre staff know the result but the participating diagnostic laboratories do not. Since it is not easy to source and prepare suitable specimens, laboratories from many countries around the world now participate in UK NEQAS schemes; the only restriction is that it must be possible to send specimens to the laboratory by post. Each participating laboratory is allocated a code which is used by the NEQAS centre to identify it in all correspondence.

The best way to use NEQAS schemes to help monitor quality in the laboratory is to treat their samples in exactly the same way as any other routine clinical specimens. In practice, this is difficult, because they usually have a distinctive packaging which is difficult to disguise, so staff often make an extra effort when testing them. In some laboratories, senior staff perform the tests or check the results, which is rather counter-productive as it this not a test of the effectiveness of normal laboratory procedures. The laboratories usually have 3–4 weeks to perform the tests they consider necessary on each sample in the set and send the results to the NEQAS centre. When the results are sent in, the reference centre allocates a score to each laboratory, based on the accuracy of the results. Each participating laboratory is then sent a report which includes the intended results, their score and the areas where they lost marks if they did not achieve the maximum possible score. They are also provided with laboratory's cumulative total score on NEQAS tests for that set of specimens (for example, HIV antibody testing) over the previous 6–12 months and an indication of how that compares with the national average (provided that more than 10 laboratories in that country are participants in that scheme).

If the laboratory has not received full marks, then the Quality Manager and the staff in that section are advised to investigate the possible reasons. It is usually possible to request a repeat aliquot of a particular sample to test again.

Sometimes the NEQAS result highlights poor practice in a particular area of the laboratory which had not been noticed previously. Other reasons for not achieving maximum marks include using a test method which is not sufficiently sensitive or because the full range of tests are not available in that laboratory, so the marks are lost for an omission rather than a mistake. For example, if the laboratory did not have the facilities to type Herpes Simplex Viruses, so sent a report of 'Herpes Simplex Virus detected' instead of the more complete 'Herpes Simplex Virus Type 2 detected', it might lose marks. The laboratory may have taken a decision not to offer typing 'in-house', but to refer samples to another laboratory for typing and only if it is considered to be clinically necessary; if their reasons are justifiable and accepted by service users, then they do not need to take any further action over the NEQAS score. Occasionally, the prepared sample deteriorates in transit and the intended analyte is not detectable by any laboratory; in this case, the results for that sample are not scored.

The cumulative score is a good indication of the quality of a laboratory's work and so departments which consistently perform poorly can be offered help to improve. Staff in the NEQAS centre cannot identify individual laboratories as they only know the codes, so they pass the information on to a separate committee called the Joint Working Group (JWG) for that particular discipline. Suitable members of this group will arrange to visit the laboratory in question and work with the staff there to try to find out why they are unable to achieve satisfactory results for that type of specimen. For example, in the 1990s, several general microbiology departments had difficulty identifying faecal parasites in NEQAS samples; investigations by the JWG attributed this to limited experience with these organisms and a lack of training opportunities for laboratory staff, rather than basic problems with laboratory procedures. This led to the development of the NEQAS-associated Parasitology Teaching Scheme, which illustrates how participation in NEQAS schemes can be supportive and educational for biomedical scientists.

4.6 Quality audit

An audit is a thorough examination of procedures in a particular situation, to make sure that minimum standards of practice are being observed. For example, when an accountant carries out inspections of all income, expenditure (and tax returns!) for a company, this would be a financial audit. Sometimes audits are done because a mistake has been noticed, but they should also be done routinely to check that there are no problems with the systems in place.

Clinical audit is the process of looking at procedures in place for all aspects of the patient's diagnosis, treatment and care. The aim in this case is to ensure that all resources are adequate and used to the full extent and that the patient has the best outcome possible (as appropriate to their condition). Biomedical scientists can be involved in audits in a number of areas within pathology, including training, safety and quality. They may also be asked to contribute to audits of an overall clinical area, such as diabetes care, as part of a multidisciplinary team.

Audits can be very useful for biomedical scientists, as they can highlight errors which have been overlooked. For example. if people were preparing up to three times the required volume of a reagent for a particular assay which had to be diluted freshly each time, the results would not have been affected. Hence the inaccuracy would not be immediately obvious, but should be shown up by an audit. If there is something wrong with a procedure, but it is not clear at which stage the problem is occurring, an audit can help to find it. In the case of the reagent where too much was being prepared for the number of tests carried out (so two-thirds of each batch was being discarded unused), the Laboratory Manager might have looked at the budget and noticed that the department was spending an unexpectedly large amount of money on that reagent. The findings of the audit would have explained why this was happening. However, audits are not always done to investigate errors, they can also be done to check that a system is actually working as well as it seems to be. If it is, then the outcome of the audit would be positive, which shows that an audit is not supposed to be simply a tool for finding malfunctions in the laboratory procedure and then choosing someone to blame for them!

There are three main types of audit which can usefully be conducted in a pathology department, 'horizontal', 'vertical' and 'examination' audit:

- *Horizontal audit:* this involves looking in detail at one stage in the sample processing within the laboratory, such as loading blood tubes on to an analyser. It entails observing how a certain (predetermined) number of specimens are handled at that stage and checking that all documents relating to that stage (e.g. SOPs, health and safety policy, training records) are relevant and up to date. It also requires ascertaining that all staff doing this task have received adequate training and are supervised appropriately.

- *Vertical audit:* this involves following the fate of a single sample through all the processing stages from specimen reception through to reporting of the result. Again, the auditor needs to observe how the specimen is dealt with at each stage, checking that all documentation is correct and making sure all staff involved are properly trained and supervised.

- *Examination audit:* this involves testing the competency of staff in a particular procedure, by watching them perform the task and again checking that written records are satisfactory.

The most common form of audit used to examine quality issues is the vertical audit, since this means following a specimen through the laboratory, as it undergoes a range of pre-analytical, analytical and post-analytical procedures. This is clearly the most helpful part of QA (see above) as it gives an overview of how the whole laboratory works.

There are set formats available for vertical audits, such as the one suggested by Clinical Pathology Accreditation Ltd (see Section 4.9), but it can be useful to think of a series of questions to ask and then decide how to answer them. Table 4.3 gives some suggestions, although this is not an exhaustive list, merely some ideas to start with; note that it is broken up into the three stages.

An audit is most effective when it is carried out objectively and meticulously – which is the approach that biomedical scientists should take to all of

Table 4.3 Suggested questions to use as part of a vertical audit

Question	Suggested evidence to answer question
Pre-analytical stage	
Was the right specimen collected from the correct patient at an appropriate time?	Check that date and time of collection are written on the specimen container and that details on container and request form correspond
Was the specimen transported to the laboratory safely and in good time?	Check the integrity of the specimen on arrival in the laboratory and date and time of collection
Is there an efficient and effective system for 'checking in' samples?	Check that the SOP for this is clear and followed by all staff in specimen reception
Analytical stage	
Is the sample suitable for testing? (has it been stored and prepared properly?)	Check the integrity of the specimen on arrival at the testing bench and consult the SOP to ensure that any special requirements have been met
Are suitable testing methods being used for this investigation?	Discuss the validity of the methods used with experienced staff and search the literature to see whether a more up-to-date test has been evaluated and is routinely used elsewhere
Are controls run regularly?	Inspect the IQC records
Is the equipment working well and safely?	Check that all equipment has evidence of being tested for electrical safety and inspect the daily maintenance records
Does everybody involved in specimen processing understand their roles?	Talk to all the staff involved and check that they have all signed to say that have read and understood the relevant SOPs
Is everybody suitably qualified, trained and supervised?	Check the qualifications, registration and training records of all staff
Is the department well managed?	Ask members of staff about their experiences of working in the department, inspect appraisal records and verify that all staff are offered opportunities to participate in continuing professional development

Table 4.3 (*Continued*)

Question	Suggested evidence to answer question
Post-analytical stage	
Are results reported by authorized staff following clear procedures?	Ensure that the SOP is clear and that all staff are following it (e.g. by asking them how results are reported)
Are telephoned results recorded?	Ascertain that records of telephoned results are kept
Are the written results well laid out?	Inspect the format of the result on the computer screen and in a printed version for clarity and ease of finding relevant information
Is it easy to find a stored sample?	Attempt to find a stored sample using the information available (e.g. sample number, plus plan of storage area)
Is clinical waste disposed of legally and safely?	Discuss procedures with staff (e.g. the safety officer) and check records of disposal

their work. However, if there are any good or bad points which the audit highlights, these must be communicated to all relevant members of staff, otherwise the audit exercise will have been in vain. It is a good idea to produce a written account of the audit findings and make them available in a public place (e.g. an electronic copy on the Trust intranet and/or a printed copy on the departmental notice board) and also to discuss them at a suitable staff meeting. Staff must work together to suggest viable changes where necessary, so that the quality of the laboratory's service can be continually improved, for the benefit of users and patients.

4.7 Clinical governance

The concept of 'clinical governance' in healthcare is analogous to 'corporate governance', which was set up to regulate the financial running of privately owned organizations. The latter is intended to protect the interests of shareholders and prevent the loss of companies' funds through fraudulent practices. Similarly, clinical governance is a system of regulation and monitoring to ensure that all healthcare organizations provide the highest quality patient care and work to national standards of 'best practice'.

Clinical governance was introduced into the United Kingdom National Health Service during the late 1990s, partly in response to concerns raised by patients and the general public about the lack of accountability of healthcare professionals. For example, at Alder Hey Children's Hospital in Liverpool, a consultant pathologist retained and stored tissue, for long-term research, which had been taken from children with serious illnesses, in many cases during the *post mortem* examination. Specific consent was not sought from the relatives of the sick or dead children and to them it seemed as though their dignity and feelings had been callously disregarded. Specific legislation was subsequently passed in response to this and other similar occurrences involving the retention of tissue (the Human Tissue Act 2004 – see www.hta.gov.uk). However, it is envisaged that clinical governance systems would reduce the likelihood of such an event, by raising the awareness of patients' rights and also because the monitoring procedures should allow early identification of any irregularities.

The underlying principles of clinical governance are that healthcare organizations and their employees should be accountable to each other and their patients, that all staff should work with honesty and integrity and that all procedures should be open and available for independent scrutiny. For staff in pathology, this means that the professional registration (see Chapter 1) and continuing professional development of individual practitioners (see Chapter 8), external accreditation of departments (see Section 4.9) and participation in internal and external quality assurance schemes (see Section 4.5.2) can be considered as part of clinical governance. However, as it is intended to be the basis of the monitoring and evaluation of overall patient care, biomedical scientists must be aware that clinical governance procedures include ensuring that practices in pathology form part of an overall high-quality experience for the patient. This means that pathology staff must liaise with staff in other departments of the hospital and primary care practitioners in interprofessional and multidisciplinary teams and participate in clinical audit (see Section 4.6), as required. Box 4.5 gives a hypothetical example of how the clinical governance system encompasses the overall patient care process.

Box 4.5 Use of clinical governance to investigate a complaint from a patient

Mrs A is 45 years old and married with two children. She noticed a breast lump and went to her general practitioner, who is part of a practice funded by Primary Care Trust B. The doctor examined Mrs A's lump and then said he would refer her to the Breast Clinic at Hospital C, which is part of Hospital Trust C.

Four months later, the Chief Executive of Hospital Trust C receives a letter from Mrs A stating that she has just received a diagnosis of breast cancer and complaining that she feels it has taken an unacceptably long time to reach this point. She also states that she cannot fault the care given to her by all the healthcare professionals that she

Box 4.5   (Continued)

has encountered during the time between noticing the lump and being given the diagnosis and she feels that it is the 'system' which has let her down.

According to the principles of clinical governance, the Chief Executive must investigate this occurrence, since the healthcare organization (Hospital Trust C) has to be accountable to its patients. He asks a senior consultant to conduct a clinical audit of how Mrs A's case was dealt with in the hospital. Biomedical scientists and their medical colleagues in the histopathology department are involved in this audit by looking in detail at the quality of their laboratory testing procedures and their turnaround times for results. Staff in radiography also undertake a similar exercise in their department, again looking at the accuracy of their diagnoses and waiting times for X-rays to be taken, and also how long it takes for results to be available. The amount of time patients are usually waiting to be seen by the consultant in the breast clinic between investigations is also considered. The qualifications, training records and registration for all staff involved are verified. The Chief Executive also asks the investigating consultant to explore the financial situation in the breast care pathway, since delays may have been caused by lack of money to employ enough staff or buy the most appropriate equipment.

This case also concerns another organization, namely Primary Care Trust B, which is asked to nominate an independent person to inspect its records. They have to check the length of time taken to refer Mrs A to the breast clinic, and also the qualifications and registration of their staff and the amount of training doctors and nurses have received in identifying potentially malignant breast lumps.

Mrs A and her family are kept fully informed of the proceedings and receive a copy of the final report, which concludes that there was a problem with the financial arrangement between Primary Care Trust B and Hospital Trust C, which meant that the initial referral from the general practitioner to the breast clinic was delayed.

4.8 Quality management system (QMS)

Maintaining and monitoring a high-quality pathology service in the 21st century is clearly complex, so it is important to have a way of coordinating everything. This is a Quality Management System (QMS) and, although it might vary in format between laboratories, it is essentially a series of documents cataloguing all the parts of the laboratory work relating to service quality. The scope has grown in recent years to meet the requirements of external accreditation (see Section 4.9). Hence it should contain records of all the IQC systems, all IQA exercises, results from all NEQAS assignments and outcomes of all audits within a specified period of time (e.g. 5 years). There should also be a timetable for internal IQA and audit, to ensure that each area of work is examined regularly (e.g. annually). Minutes of all meetings where quality issues were discussed (general staff meetings, managers' meetings) must be kept and they should be available for staff to read. A plan for the skill mix of staff within a department, each person's qualifications and their training records should also be included in the QMS documentation. SOPs (see Chapter 5) are controlled documents, which means that only authorized staff members can change them and they have to be regularly reviewed. The QMS should include the master copy of all SOPs in a format which precludes unauthorized access (for example, in a password-protected electronic form). The QMS needs to be organized and administered by the Quality Manager or Quality Officer, who can remind staff in a particular section that an audit of their area is due and can make it a priority to follow up any problems relating to quality which arise, while other staff continue with their routine work. In this way, thinking about quality becomes a normal part of all biomedical scientists' working days, without being a distraction from service delivery.

4.9 Accreditation

Accreditation is official recognition that someone or something has met an approved standard. It is necessary for two main reasons, the first being to ensure that companies and organizations providing a service to others do meet minimum standards of quality, reliability and safety. This should allow potential customers to make judgements about whether they want to use that service. The second purpose of accreditation is to have a means to call to account those companies and organizations which do not meet minimum standards. In many accreditation schemes, companies have to apply for re-accreditation regularly and this can be refused if practices have changed and standards have fallen. In the United Kingdom there is one scheme which applies to all clinical laboratories, which is that run by Clinical Pathology Accreditation Ltd (CPA). There are also programmes organized for specific areas of work within clinical pathology. These include the United Kingdom Accreditation Service (UKAS) scheme for work in Public Health Laboratories, the Medicines and Healthcare Products Regulatory Agency (MHRA) programme to oversee standards in blood transfusion services and the Human Tissue Authority (HTA) arrangements to ensure that histopathology departments comply with the law on handling and retaining human tissue. They all work in close partnership to avoid duplicating work and to keep the accreditation process as simple as possible for laboratories.

The CPA was specifically set up in 1992 to provide a means of peer review of clinical pathology laboratories and the NEQASs in the UK. Using the expertise of a panel of experienced biomedical scientists and consultant pathologists in each pathology discipline, a set of standards was created by which laboratories could be judged on the quality of their pathology service. Laboratories who apply for accreditation must have written evidence that they meet all the standards and the department is inspected by a medical consultant and a senior scientist (biomedical scientist or clinical scientist) who work in another laboratory but in the same discipline. As more laboratories achieved accredited status, it became an accepted mark of high professional standards, although applying for CPA accreditation is voluntary. It is now unusual, 15 years later, to find a clinical pathology department that does not have or is not working towards accreditation with CPA.

The original CPA standards were revised in 2001–02 and now fall into eight categories:

A. Organization and quality management system
B. Personnel
C. Premises and environment
D. Equipment, information systems and materials
E. Pre-examination process*
F. Examination process*
G. Post-examination process*
H. Evaluation and quality assurance

(*in other parts of this chapter, the terms 'pre-analytical phase', 'analytical phase' and post-analytical phase are used to mean the same as the 'pre-examination process', 'examination process' and 'post-examination process', respectively. given here).

Careful inspection of these categories will show that all the aspects of quality mentioned elsewhere in this chapter are included in the list.

Each of these categories contains sub-categories, which are given in Table 4.4. All of these must be addressed in the documents which the laboratory submits for inspection. The quality issues and measures discussed within this chapter are all addressed in these standards. For example, A7 requires a laboratory which is applying for CPA accreditation to have designated a member of staff as the 'Quality Manager' (see Section 4.2), while B9 specifies that there should be opportunities for education and training for all staff and this must be recorded (see Section 4.4). Standards E–F focus on aspects of the pre-analytical, analytical and post-analytical phases discussed in Section 4.4, which highlights how important they are to service quality. To meet standard H, a laboratory must have evidence of Internal Quality Control, regular Internal Quality

Table 4.4 Details of CPA standards

Category	Sub-category
A: Organization and quality management system	A1: Organization and management A2: Needs and requirements of users A3: Quality policy A4: Quality management system A5:Quality objectives and plans A6: Quality manual A7: Quality Manager A8: Document control A9: Control of process and quality records A10: Control of clinical material A11: Management review
B: Personnel	B1: Professional direction B2: Staffing B3: Personnel management B4: Staff orientation and induction B5: Job descriptions and contracts B6: Staff records B7: Staff annual joint review B8: Staff meetings B9: Staff training and education
C: Premises and environment	C1: Premises and environment C2: Facilities for staff C3: Facilities for patients C4: Facilities for storage C5: Health and safety
D: Equipment, information systems and materials	D1: Procurement and management of equipment D2: Management of data and information D3:Management of materials
E: Pre-examination process	E1: Information for users and patients E2: Request form E3: Specimen collection and handling E4: Specimen transportation E5: Specimen reception E6: Referral to other laboratories
F: Examination process	F1: Selection and validation of examination procedures F2: Examination procedures F3: Assuring the quality of examinations
G: Post-examination process	G1: Reporting results G2: The report G3: The telephoned report G4: The amended report G5: Clinical advice and interpretation

Table 4.4 (*Continued*)

Category	Sub-category
H: Evaluation and quality assurance	H1: Evaluation and improvement processes
	H2: Assessment of user satisfaction and complaints
	H3: Internal audit of quality management system
	H4: Internal audit of examination processes
	H5: External quality assessment
	H6: Quality improvement

Taken from www.cpa-uk.org.uk.

Assessment and audit exercises and participation in External Quality Assessment (i.e. NEQAS), which are all mentioned elsewhere in this chapter. The physical features of the environment and the general running of the department which are considered by standards C and D include some of the key factors which affect the quality of a department's service referred to in Section 4.10.

The CPA inspectors will look for evidence that the laboratory applying for accreditation meets the minimum standards in all these areas. They are particularly vigilant to check that the relevant procedures are in place at all times – not just in the few weeks leading up to the inspection visit!

After the visit, the inspectors prepare a report which will recommend one of three outcomes:

- Full Accreditation, which means that the department is deemed to have met the standards in all areas. They would also propose a re-inspection within a defined period of time (e.g. 3 years)

- Conditional Accreditation, which means that they have found satisfactory evidence that most of the standards have been met, but they are not happy about procedures in a few areas. This is the most common outcome of an inspection and departments are given a set amount of time (e.g. 3 months) to address the problem and submit relevant documents to the CPA. A re-inspection is not always necessary, but it depends on the individual case.

- Non-accreditation, which means that the inspectors found serious deficiencies within the department and they feel that conditional accreditation with re-inspection would not be enough to address them. In this case, the CPA would work with that department to tackle the problems and raise standards and ask them to re-apply for a new inspection once this has been achieved.

Preparing for a CPA inspection and being present on the day of the visit can be a very stressful process for laboratory staff. Most departments view CPA accreditation as a marker of high quality and want to do well. It is hard work to document everything which is done in the laboratory towards providing a high-quality service (and it requires staff to take time away from the bench and day-to-day service provision), but it is worth the effort. It shows the staff that their work is of a high standard and it also gives confidence to the users of the service.

4.10 Factors affecting the quality of work in a pathology laboratory

There are a number of factors which can affect the quality of the service which a particular pathology laboratory provides. Even in laboratories with the most rigorous quality management systems, mistakes can occur and these are often due to human error. For example, if a biomedical scientist is feeling under pressure to provide a result quickly and the assay has given a credible answer for the test sample results, although the internal quality control indicates a breach of one of Westgard's rules, they might be tempted to report the result anyway. The biomedical scientist ought to investigate the causes of the breach and re-run the assay, but sometimes people feel stressed and unsupported and so make poor decisions. Another reason for incorrect results being reported is that despite being aware that they might have made a mistake in conducting the test, the biomedical scientist is afraid to report it to a senior colleague, so the erroneous result is let through the system. When running low on a reagent, it is sometimes possible to dilute it and override the conditions set for the assay which produces a result; this saves money and could also reduce delays in turnaround times when the reagent has not been ordered in time. None of these are advisable courses of action and they are all examples of unprofessional and unscientific practice, but they can happen because biomedical scientists are human beings.

Hence the quality of a pathology service is not only affected by the suitability and rigour of the quality management system. Diagnostic laboratories are generally 'closed' environments, meaning that the same small group of people work together every day in the same room. Individual personalities and mood changes can influence how people feel about working in that laboratory, as much as the job itself or the work load that day. The way in which the department is managed has a very important influence. Where staff feel that their supervisor and the Laboratory Manager value their work and support them, they are more likely to work better and to admit to mistakes, confident that they will not be blamed in a personal way; the problem can then be remedied before an inaccurate result is sent out. This 'no blame' environment encourages staff to be open about errors and problems and helps to keep improving the quality of that laboratory's service. The in which way members of the laboratory team relate to each other is also a factor in service quality. Where people get on well and trust one another, the tasks will be completed more efficiently and, in times of particular stress, colleagues are more likely work through them as a team. Since biomedical scientists and other laboratory staff often spend long periods of time in one room, the physical environment can affect their work. A room, in an old building, which was not designed to hold large items such as analysers, away from the main hospital, can feel very cluttered and depressing. By contrast, staff working in a modern purpose-built laboratory block included on the main hospital site and containing all the latest equipment can feel motivated to fulfil their tasks well; they would be more likely to feel valued as scientists and to remember their role in patient care.

Quick quiz

1. Why is the quality of the laboratory service important in clinical pathology?

2. Give definitions for five of the seven key terms relating to quality in clinical pathology departments.

3. What is the main role of the Quality Officer or Quality Manager in a clinical laboratory?

4. How is Internal Quality Control (IQC) for staining of slides carried out?

5. Explain the purpose of Westgard's rules in IQC.

6. Distinguish between 'systematic' and 'random' error in laboratory procedures.

7. What is the difference between quality control and quality assurance?

8. Why is the monitoring of laboratory equipment an important aspect of quality assurance?

9. Give two reasons why selecting a suitable format for the laboratory report is vital to service quality.

10. How can all biomedical scientists contribute to quality assurance in their department?

11. What is internal quality assessment?

12. Approximately how many schemes are part of UK NEQAS?

13. Name three main discipline group within NEQAS.

14. Which members of the laboratory staff should handle NEQAS specimens?

15. What is the role of the Joint Working Group in each NEQAS scheme?

16. Give a definition of a clinical audit.

17. Which of the three types of audit used in clinical pathology laboratories is most useful for monitoring quality?

18. What is the 'post-analytical phase' of specimen processing?

19. How can the results of an audit be disseminated to laboratory staff?

20. Why was clinical governance introduced into healthcare organizations in the United Kingdom?

21. Name four areas of the laboratory activity which should be documented in the Quality Management System.

22. What is the purpose of laboratory accreditation?

23. How can management style affect the quality of work in a clinical laboratory?

Suggested exercises

1. *IQC exercise*

   In Laboratory B at Hospital Y, the Rubella antibody assay is run every weekday in an automated serology system. The reagent kit includes a high positive, a low positive and a negative control, each of which must be included each time the test is run. The criteria set out by the kit's manufacturer for accepting the assay *each time it is run* are:

   (a) high positive control $> 3\times$ negative control

   (b) low positive control ≥ 2

   (c) negative control.

   In addition to using the manufacturer's controls, Laboratory B includes its own 'in-house' internal quality control (IQC) sample, made up from pooled serum. Before being introduced as a control, this pooled serum control was tested in the Rubella antibody assay 20 times. This allowed the following information to be determined:

   Mean antibody titre: 15.0 IU/mL, standard deviation 0.3.

   Table 4.5 records the results from the routine running of the kit controls and the in-house IQC sample during December 2006.

Table 4.5   Results obtained in the Rubella antibody assay run in Laboratory B at Hospital Y during December 2006

Date	High positive (IU/mL)	Low positive (IU/mL)	Negative (IU/mL)	IQC sample (IU/mL)
1.12.06	25.3	11.1	5.3	15.0
4.12.06	27.1	12.3	4.7	14.9
5.12.06	25.7	11.3	5.5	15.3
6.12.06	26.4	12.1	4.5	14.5
7.12.06	27.1	10.9	5.1	15.1
8.12.06	27.2	11.5	5.3	15.4
11.12.06	25.4	12.1	5.1	14.8
12.12.06	26.5	11.2	4.9	15.2
13.12.06	26.1	11.9	5.2	14.9
14.12.06	25.9	11.2	5.6	15.3
15.12.06	25.4	11.7	4.8	14.9
18.12.06	25.3	11.4	4.9	14.8
19.12.06	25.0	12.1	4.9	14.9
20.12.06	25.2	11.3	4.9	14.8
21.12.06	24.6	11.8	4.8	14.7
22.12.06	24.5	11.1	4.8	14.6
27.12.06	24.6	11.4	4.7	14.7
28.12.06	24.5	11.9	4.5	14.6
29.12.06	24.5	11.5	4.5	14.5
30.12.06	24.3	11.1	4.3	14.3

Use these data to answer the following questions:

(i) Have any of the runs failed the kit manufacturer's criteria for accepting the assay? If so, on which dates did the controls fail and which of the criteria were not met?

(ii) Draw a plot for the 'in-house' IQC sample. Use this to apply Westgard's rules to the data.

(iii) Have any of Westgard's rules been broken in the Rubella assay during December? If so, which rules were broken and on which dates?

(iv) What are the implications of your findings for the staff running the Rubella assay in Laboratory Y?

2. *EQA exercise*

(a) Find out about all the NEQAS schemes in which the laboratory you are working in at present participates, by searching the NEQAS website for information.

(b) Obtain a copy of a printout of a recent set of NEQAS results. Photocopy it, anonymize it, then using the facts you have found out in part (a), *annotate* it to explain what *all* the information means.

3. *Audit exercise*

(a) Perform a vertical quality audit following the guidelines given by CPA.

(b) Prepare a written report of your findings, highlighting areas of good and bad practice, and recommend changes that you think could be useful to improve the quality of the service in your department.

Suggested references

Ajeneye, F. (2007). Pre-analytical quality assurance: a biomedical science perspective. *Biomedical Scientist*, **51**, 86–87.

Burnett, D. (2002). *A Practical Guide to Accreditation in Laboratory Medicine.* London: ACB Venture Publications.

Clinical Pathology Accreditation website: www.cpa-uk.co.uk.

Hewitt, J. (2005). Effective internal quality control. *Biomedical Scientist*, **49**, 26–28.

Human Tissue Authority web site: www.hta.gov.uk.

Institute of Biomedical Science, *Error logging in clinical laboratories.* www.ibms.org/pdf/error_logging.pdf.

McSherry, R., and Pearce, P. (2007). *Clinical Governance: a Guide to Implementation for Healthcare Professionals*, 2nd edn. Oxford: Blackwell.

NEQAS website: www.ukneqas.org.uk.

Pitt, S. J., and Sands, R. L. (2002). Effect of staff attitudes on quality in clinical microbiology services. *British Journal of Biomedical Science*, **59**, 69–75.

Westgard rules website: www.westgard.com.

5

Basic principles of working in a clinical pathology laboratory

5.1 Introduction

Many biomedical scientists find that employment in a clinical pathology laboratory can be very rewarding. Processing large numbers of specimens, using the highest possible standards of quality and in good time to be clinically useful, can bring enormous job satisfaction at the end of a working day. It can be equally fulfilling to spend several hours attempting to solve a technical or scientific problem with a particular analyser or reagent. Biomedical scientists are usually part of a team including colleagues from other professional groups and sometimes other pathology disciplines, which can be stimulating, (see Chapter 2). They are also continually learning through their day-to-day experiences and are required to keep up with improvements in technology and scientific advances (see Chapter 8). However, if the work is not organized well, the clinical laboratory can be a stressful and even hazardous environment. This is why there are rules and policies in place which staff are expected to learn and follow. It may seem that every aspect of work has a regulation which must not be broken. However, careful thought will show that in many cases the approach they are advocating is sensible and is intended to ensure that results are accurate and that all staff are healthy and safe while working in the laboratory. The reason for some rules which might seem 'silly' or 'petty' is probably because someone in the past showed a distinct lack of common sense and did something dangerous or made a mistake with a patient sample which affected the result. This chapter describes the stages of the passage of samples through the pathology laboratory from the patient to the reporting of results and in this context discusses specific aspects of professional behaviour regarding health and safety and confidentiality.

An Introduction to Biomedical Science in Professional and Clinical Practice   Sarah J. Pitt and James M. Cunningham
© 2009 John Wiley & Sons, Ltd

5.2 Working as a professional biomedical scientist in the laboratory

As shown in Chapter 6, a wide range of specimens are handled in pathology departments, including blood samples, urine, faeces, aspirated fluids, swabs from infected sites, pieces of tissue and whole organs. The way in which a type of sample is treated may vary in each discipline, depending on the nature of the investigation. For example, a clotted blood sample which is sent for serological testing will usually be centrifuged on arrival on the testing bench, the serum separated and an aliquot pipetted into a separate container, regardless of whether the test requested is biochemical, haematological or microbiological. However, a whole blood sample (i.e. after mixing with an anticoagulant) might be put through an analyser to test for a biochemical marker, used to prepare a blood slide to allow examination of white blood cells or incubated to culture microorganisms. Regardless of the type of specimen or the nature of the assay, all samples must be handled safely. This means that all staff must adhere to written procedures, but also listen to advice from senior colleagues in order to avoid mistakes and accidents. It is important for biomedical scientists to remember that laboratory work is team work, so they should not jeopardize the safety of colleagues by taking risks. Similarly, they should not forget the patient who awaits the information that the results will bring, and so must not compromise service quality by cutting corners.

Biomedical scientists are expected to behave professionally at all times, following codes of conduct (see Chapter 1) and fulfilling the duties agreed in their job description. In most cases, this will include the level of responsibility that the person in that post is required to take, and also who they are expected to report to as their line manager. The contract of employment will specify the number of hours per week that the biomedical scientist is paid to work, but their start and finish times must be agreed with their manager. This is to ensure that enough staff are available when the bulk of the work comes into the laboratory, in addition to dealing with the less busy times (such as early evenings) and providing cover for times when results are only required in an emergency (such as at night and weekends). It is important for biomedical scientists to realize that they are part of a team and that colleagues will be relying on them to carry out the tasks assigned to them, in order to contribute to the smooth running of the department. This means that if someone is unable to attend work, for example due to illness, they must inform the laboratory as soon as possible, so that work can be allocated to other people as necessary. It is equally important not to go to work when very infectious or against medical advice, as this can be dangerous both for the sick individual and for those around them.

Another aspect of acting as a professional is to observe the laws and guidance on confidentiality (see below). Despite rarely seeing patients, biomedical scientists have access to personal details about patients, including their date of birth, address and, of course, information about any illness or condition that might be under investigation. It is very important not to share these particulars with any one outside the hospital, for two reasons. The first is that patients provide personal information in the expectation that it will not become public knowledge. So, for example, if a sample comes into the laboratory from an acquaintance, giving their full name which includes an amusing middle name, this should not be shared with everyone at a party! More seriously, if a biomedical scientist finds out the test results for a relative or neighbour, they are not allowed to disclose them to the patient. This is because telling someone the scientific part of the result out of context can be distressing and even dangerous; the patient might need treatment and advice, which need to be given by the doctor who has overall knowledge of their case. Telling someone that their haemoglobin levels are 'low' might lead them to decide buy some iron tablets and miss the appointment with their doctor. However, there are a range of possible underlying causes for anaemia, which include malignancies, so in this case, rather then being helpful, the biomedical scientist could actually be hindering their friend's treatment.

5.3 Flow of work in a clinical laboratory

All clinical pathology laboratories are busy environments in which to work. Approximately 25 000 biomedical scientists are employed in pathology departments across the United Kingdom and they are involved in processing about 200 million samples each year. In most pathology departments, the biochemistry laboratory handles the highest number of specimens per year, followed by followed by haematology, then microbiology, while the fewest samples are sent to histopathology. However, staff in each laboratory work equally hard, since many of the tests performed for biochemical markers can be finished within an hour or two, whereas it can take several days for a tissue to be processed in histopathology. When analysing the work of a diagnostic laboratory, it is useful to think about how specimens move through the department. Regardless of the discipline or types of assay performed, the flow of work can be divided into three stages, namely the 'pre-analytical', 'analytical' and 'post-analytical' phases. Figure 5.1 shows 10 steps in the processing of a sample through a clinical laboratory. The pre-analytical phase comprises all the steps from when the test request is generated from a particular patient through to the sample being taken to the bench within the laboratory to be tested (1–4 in Figure 5.1). The preparation and testing of the sample constitutes the analytical phase, (steps 5–8) and the inspection of the output from the test assay, the recording, validation and reporting of results (steps 9–13) are the post-analytical phase. Within each phase, the tasks are clearly defined and all staff should understand their

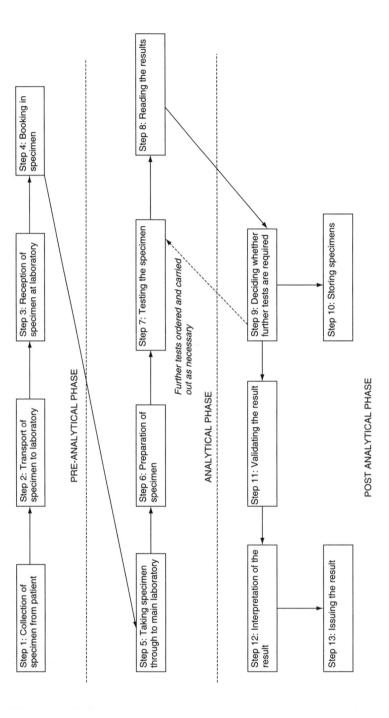

Figure 5.1 Flow chart of passage of a typical speciman through a clinical laboratory. N.B. Standard Operating Procedures written in accordance with the Health and Safety at Work Act 1974 and the Control of Substances Hazardous to Health Regulations 2002 will be available for all tasks. For step 3 onwards, these will be controlled documents managed by staff within the pathology department.

roles within the team. Some jobs can be undertaken by a person at any grade, for example, taking waste bins to be autoclaved. Other activities, such as validating results, must be restricted to suitably qualified people. All aspects of work within the laboratory should be described, in writing, in a 'standard operating procedure' (SOP) and be carried out with due consideration for the health and safety of staff.

To illustrate what information can be gathered by looking at the different steps in detail, this section will follow the progress of a particular specimen. In this case, it will be a routine blood specimen, sent to the Microbiology Department for Rubella IgG antibody testing, but the important points will be relevant to all types of sample in all disciplines.

The patient in this case is Mrs AP, who is 12 weeks pregnant and is attending the Antenatal Clinic (ANC) for her first appointment; this is usually called the 'booking appointment'. The midwife sees her in the clinic and asks about her general health and well-being and plans for the birth. She examines Mrs AP, sends her for an ultrasound scan and collects a series of blood samples known as 'booking bloods'. This is Mrs AP' s first pregnancy, but her general practitioner confirms that he is not aware of any underlying health problems. At this stage, therefore, some blood is collected into an EDTA tube and sent to haematology for full blood count, blood grouping and Rhesus testing; a second blood sample is taken and put into a lithium heparin tube to be sent to biochemistry for a random blood glucose sugar test. A third blood sample is put into a plain tube (i.e. to allow it to clot) and sent to the microbiology department, to be tested for Hepatitis B surface antigen, anti-Human Immunodeficiency Virus antibodies, *Treponema pallidum* haemagglutination and anti-Rubella IgG.

5.3.1 Pre-analytical phase

In this case, the process starts at 10 a.m. on a Tuesday morning in the ANC, where Mrs AP is seen by a midwife (Mrs JK) for her routine booking appointment (Figure 5.2). The rest of this chapter follows the blood sent to microbiology and specifically the anti-Rubella antibody test.

Step 1: Collection of specimen. Mrs JK discusses the reasons for taking blood with Mrs AP and explains why it is important to test for Rubella antibodies at an early stage in the pregnancy (to make sure that she is immune and so could not pass on congenital Rubella to her baby if she were to come into contact with the infection). Mrs JK makes sure that Mrs AP understands this and consents to the Rubella antibody test. She then completes a request form, using a preprinted sticky label to provide the 'minimum data set for patient identification' (see Box 5.1) and listing the tests requested. Before taking the blood, the midwife also carefully writes Mrs AP's name, date of birth and the time of collection on

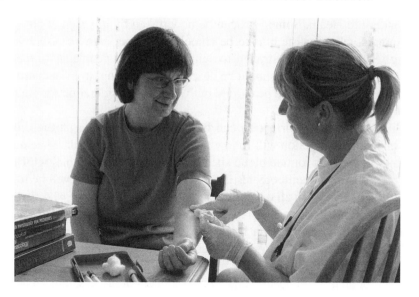

Figure 5.2   Midwife collects blood from patient Mrs AP

Box 5.1   Minimum data set for patient identification

The biomedical scientist who performs the assay on a particular patient's sample has usually not collected the specimen or indeed met the patient themselves. Therefore, laboratory staff need confirmation that the person who did take the sample selected the correct patient, to avoid issuing erroneous results. This is done through only accepting specimens which meet the requirements for the minimum data set for identification. The patient details given on the request form must correspond with those written on the specimen container.

Acceptable identification data include:

- Full name

- Date of birth

- Gender

- Hospital number or NHS number.

In most laboratories, if three of these forms of identification were given and they matched, then the specimen would be accepted for testing. If there was a minor discrepancy in one of these data sets – for example, spelling a person's name as 'Anne' on the request form and 'Ann' on the specimen container, the sample still might be accepted, provided that the surname and date of birth matched. Sometimes, a quick telephone call to the ward or clinic where the sample was collected can confirm the patient's identify (e.g. samples were only collected from one person of that name today, so it was a simple spelling error). A note would be included with the result, to the effect that despite this mis-match, pathology staff had verified the patient identity with the person who collected the sample. This approach is used particularly when the specimen is difficult to collect – for example, cerebrospinal fluid. However, if the date of birth on the request form was completely different from that on the specimen container, it would be unwise for the laboratory to assume that they related to the same patient. In this case, the ward or clinic would be informed and a fresh specimen requested. The SOP will give guidance on when to accept or reject mis-labelled specimens.

However, for blood samples sent for cross-matching prior to blood transfusion, all four forms of identification must be provided on both the request form and specimen container and they **must** match perfectly, otherwise the laboratory will not proceed with the investigations. Preprinted sticky labels are sometimes used on specimen containers, but blood for cross-matching must be handwritten (to show that the person taking the sample has made a conscious effort at getting details right). This is because the consequences of giving blood of the wrong type to a patient are so serious (potentially fatal) that the pathology staff need to be absolutely certain that the blood they are testing comes from the right patient. For information about errors in blood transfusion and strategies to avoid them, see the Serious Hazards of Transfusion website (shot.org.uk).

the side of the collection tube. This is important, so that staff in pathology can be confident that the sample comes from the correct patient and that they have consented to the tests requested (see Figure 5.3). Since the test assay requires serum, there should be written instructions available in the clinic, informing the midwife that the blood must be collected into a plain tube (i.e. one containing no anticoagulant). Contact details for pathology should be readily accessible (e.g. within the written instructions or on the hospital intranet system), in case the midwife has a query about collecting the sample. Blood is taken from Mrs AP at 10.25 a.m. on Tuesday.

Step 2: Transport of the specimen. Mrs JK checks that the blood tube is sealed and places it in a special clear plastic bag which has two compartments – one for the sample itself, which can be folded over to make an enclosed section, and the other for the request form. This is in case the blood bottle is broken

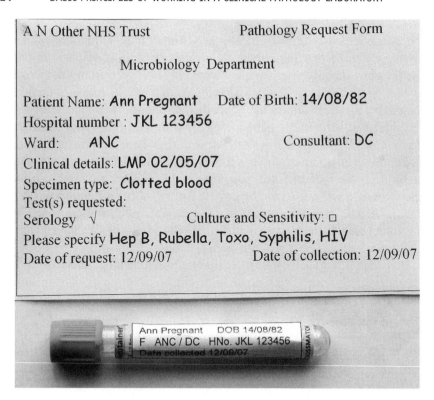

A N Other NHS Trust   Pathology Request Form

Microbiology Department

Patient Name: **Ann Pregnant**   Date of Birth: **14/08/82**
Hospital number : **JKL 123456**
Ward: **ANC**   Consultant: **DC**
Clinical details: **LMP 02/05/07**
Specimen type: **Clotted blood**
Test(s) requested:
Serology √   Culture and Sensitivity: □
Please specify **Hep B, Rubella, Toxo, Syphilis, HIV**
Date of request: 12/09/07   Date of collection: 12/09/07

Ann Pregnant   DOB 14/08/82
F ANC / DC HNo. JKL 123456
Date collected 12/09/07

Figure 5.3 Specimen request form and blood tube for Mrs AP, showing that patient details match

or becomes leaky *en route* to the laboratory. The spillage will be contained in the enclosed section of the bag, while the request form will remain intact and the laboratory staff can use the information on it to contact the ANC to ask for another sample. Since this blood sample has been collected in the morning, it will join the batch of samples from that ANC session to be collected by a porter at 12.30 p.m., so it will arrive in the laboratory within a few hours of being taken. If there was likely to be a delay, the blood should be stored at +4 °C until it can be taken to the Pathology Department. If blood is left at room temperature overnight, the red blood cells start to lyse and this can affect the reliability of the results in many assays. The porter (Mr DH) has been provided with a strong, lockable container to carry the specimens from the ANC to pathology. He arrives at pathology Specimen Reception at 1.10 p.m. and informs the person who is receiving specimens that he has brought samples from the ANC and empties his container into the box provided on the counter. He is given a set of printed reports of results from previous tests to return to the ANC.

Step 3: Reception of specimen. Once the sample has arrived in pathology, everything that any member of staff does with it will be directed by a SOP (see

Box 5.2), taking into account the Health and Safety at Work Act (see Box 5.3) and the Control of Substances Hazardous to Health Regulations (see Box 5.4). The Medical Laboratory Assistant who is working in Specimen Reception (Mr CF) then begins the task of checking through all the samples which have just arrived. As required by the SOP, he is wearing appropriate personal protective equipment (PPE), which is a laboratory coat and gloves. Mr CF takes each specimen bag in turn and checks that the sample bottle has not broken or leaked into the plastic bag before taking out each blood sample and placing it in a rack. If one of the samples has leaked, then there should be an SOP to guide staff in safe handling of the sample and contacting the ward or clinic to explain the problem and request a fresh specimen. He also removes the request form and reads it carefully, to make sure that the patient's name, date of birth and hospital number are clearly shown, along with the tests requested. Mr CF then ascertains that the name and hospital number on the blood bottle match those given on the request form. As Mrs AP's sample has arrived intact and all the identifying details match, it can be passed on to the next stage.

Box 5.2 Standard operating procedures

It is now usual to have written guidelines for all tasks undertaken in the laboratory. These are called standard operating procedures (SOPs) and they cover everything from how to do the most complicated laboratory methods, to how to communicate an urgent result by telephone, to when and where a laboratory coat must be worn. These SOPs are written by a senior member of staff working in the relevant section of the laboratory and include health and safety considerations and which grades of staff can carry out the procedure, in addition to the detailed method for performing the task.

When a member of staff starts working on a particular bench, they must be fully trained in all the tasks they will be required to undertake. This training includes reading and demonstrating understanding of, all the relevant SOPs and signing a document to say that they have done this. This should ensure consistency in the way that particular tests are performed, irrespective of the individuals who did the work. To make sure that everyone follows the same procedure, all SOPs are designated as 'controlled documents'. This means that only the original author (or another authorized person) is allowed to change anything in the document. If a biomedical scientist notices that an alteration to the method seems to be necessary, they should not write the new method on the SOP, but should discuss it with the document's controller. This stops little changes creeping in over time, which may eventually affect the quality of the results.

Box 5.3   The Health and Safety at Work Act 1974

Under the terms of the Health and Safety at Work Act 1974, an **employer** must ensure that:

- The workplace is a safe as possible and minimize the risk to employees from their duties at work.

- Written procedures regarding health and safety are in place (and regularly revised).

- Employees are trained and supervised adequately so that they do not jeopardize their health and safety or that of others while at work.

- There are robust procedures for working with employees' representatives (e.g. union representatives) and outside agencies on matters regarding health and safety.

- Employees' health is monitored (particularly when exposure to a health risk is inevitable, e.g. people who work with radioactivity).

- Visitors to the premises are protected from risk (e.g. advised of procedures, not left unattended),

It is the responsibility of **employees** to ensure that they:

- Take reasonable care with regard to their own health and safety and that of other people in the workplace.

- Cooperate with their employer over matters of health and safety.

Box 5.4   Control of Substances Hazardous to Health Regulations 2002 (COSHH)

Under these regulations, employers are required to assess the risks and minimize employees' exposure to substances which can be harmful. In this context, a 'substance' includes everything which might be immediately lethal, can cause harm after prolonged exposure or which may be hazardous to some people. Thus biological agents such as HIV virus, liquid chemicals such as formalin and latex within gloves are all substances regulated by COSHH.

Employers are required to:

1. Assess the risks for all substances used in the workplace.

2. Decide what the appropriate precautions are for each substance.

3. Implement control measures to prevent employees from being exposed to the substance or limit the exposure as far as possible.

4. Make sure that these control measures are always used.

5. Monitor employees' exposure to hazardous substances where necessary.

6. If necessary, give staff regular health checks.

7. Have plans in place to deal with accidents, incidents and emergencies in case they occur.

8. Make sure that employees are aware of the dangers from hazardous substances, have full training in handling them and are properly supervised.

In clinical laboratories, the employer is the hospital or NHS Trust, but the duty of risk assessment of individual substances is usually delegated to laboratory managers and biomedical scientists in senior grades. All SOPs must include assessment of the risks to staff carrying out the procedure in accordance with COSHH; hence this is the duty of the person who writes the SOP and is the document controller.

Step 4: Booking in the specimen. The request forms are taken to a clean office area where an administrative assistant (Mrs EM) enters the patient details, the date and type of specimen received and the tests requested, on the computer system. There will be an SOP outlining the correct procedure for this which Mrs EM will have signed to say that she has read and understood it. The computer system is regularly updated and there are training sessions for staff to make sure that they are aware of the changes. The entry of Mrs AP's data on to the system allows the generation of a unique laboratory number, by which the sample can be identified on its journey through the laboratory. In the case of Mrs AP's blood sample for viral serology, it is allocated the number S0710214. A label is printed with this number on, which Mrs EM puts on to the request form. Some smaller labels are also made – one for the original blood sample and the other to go on to a 2 mL vial (see step 6); these are attached to the request form with a paper-clip. Once she has completed a batch of request forms, Mrs EM takes them back to the specimen reception area and puts them in a designated box, next to the rack of blood tubes which carry the same laboratory numbers. Mr CF puts the sticker with the laboratory number on the appropriate sample container and the specimens are left to be collected by a member of staff from the appropriate laboratory – in this case microbiology.

5.3.2 Analytical phase

Step 5: Taking specimen through to main laboratory. At 2.05 p.m., a Medical Laboratory Assistant from the microbiology department (Miss DL) comes into specimen reception to collect all the samples which have been received for microbiological investigations – including Mrs AP's blood sample. Miss DL is wearing PPE (and laboratory coat and gloves) and puts all the specimens and request forms into a box which has a handle, to allow them to be carried safely. She takes them through to the sorting and preparation area within the microbiology laboratory.

Step 6: Preparation of the specimen. Miss DL takes each specimen in turn and checks that the patient identification details on the request form (which now include the laboratory number) match those on the specimen container. She also examines the samples to ensure that they do not appear to have deteriorated, which would mean that it was not possible to test them or that any test results would be unreliable. The blood specimens which need to be centrifuged, so that serum can be separated from them, are lined up in the rack, in order of their laboratory number. Miss DL then carefully labels a 2 mL vial for each sample, using the preprinted stickers which are with each request form and puts them in numerical order in another rack (Figure 5.4). The blood bottles, which include Mrs AP's sample, are placed in a holder inside a centrifuge bucket, ensuring that volumes are balanced (see Chapter 7). Once they have been spun down, Miss DL carefully and methodically aspirates the serum from each blood sample into the correctly labelled 2 mL vial. While doing this, she wears laboratory coat, gloves and eye protection. The filled serum vials are placed in numerical order in a larger rack and stored at $+4\,°C$ until they can be tested. Mrs AP's serum sample is placed in the refrigerator in the microbiology department at 3.50 p.m. on Tuesday. The blood tubes are also stored in the refrigerator for a week before being disposed of safely. From Miss DL's point of view, this means putting the blood bottles into a waste bag which indicates that the contents may present a biohazard (usually yellow). This is autoclaved to destroy any infectious agents, before being taken away from the hospital site for disposal.

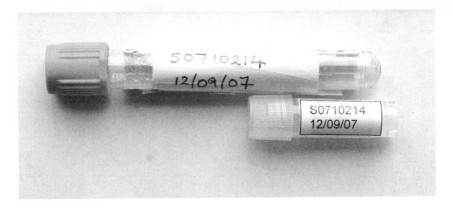

Figure 5.4 Blood tube and serum vial both showing the allocated laboratory number S0710214

Step 7: Testing the specimen. The routine anti-Rubella IgG assay is run once a week, on a Thursday. Therefore, Mrs AP's serum sample is stored in the laboratory refrigerator for 42 hours, until 9.50 a.m. on Thursday, when it is taken out ready for testing by Mr FM, one of the Specialist Biomedical Scientists. Prior to organizing the serum samples, Mr FM has used the computer system to generate a 'work list' containing the laboratory numbers of all the samples which require anti-Rubella IgG testing that day (Figure 5.5). Wearing PPE and following the appropriate SOPs, he carries out the test procedure. He finds all the serum vials labelled with the numbers corresponding to those on his list and places them in a rack. He then takes them to the work bench and leaves them to warm to room temperature while he prepares the reagents and equipment needed for the test. The anti-Rubella IgG assay used in this laboratory is an enzyme-linked immunoassay (EIA), which comes as a commercial kit, where the solid phase is in the form of a 96-well plate separated into strips and each well is coated with Rubella antigen. The test is performed using a semi-automated system, which involves using a piece of equipment to process the EIA. This contains a robot arm to pipette the required volume of test serum or reagent into each well during the assay and has a heated block to allow incubation at the correct temperature according to pre-set programmes, but the reagents have to be prepared in advance (e.g. diluted) by the operator.

Mr FM ascertains how many test wells he will need for this run of the assay, by counting the number of test samples then adding the controls and internal quality control sample (see Chapter 4). He then calculates the volume of each of the reagents required for the test run and the amount of wash buffer which should be available and prepares these. He then places the test serum and control vials in the processor, being careful to ensure that they are in the order in which they appear on his work list. He removes the lids from the vials, sets the machine to the appropriate programme and starts the run at 10.20 a.m.. Mr FM spends

Serology Work sheet:		Rubella IgG		14/09/07
Well number	Sample ID	Well number		Sample ID
1	Blank	25		S 0710176
2	Neg control	26		S 0710177
3	Neg control	27		S 0710178
4	Neg control	28		S 0710190
5	High Pos control	29		S 0710191
6	High Pos control	30		S 0710192
7	High Pos control	31		S 0710193
8	Low Pos control	32		S 0710194
9	Low Pos control	33		S 0710195
10	Low Pos control	34		S 0710196
11	IQC 726	35		S 0710197
12	S 0710148	36		S 0710198
13	S 0710149	37		S 0710199
14	S 0710150	38		S 0710200
15	S 0710151	39		S 0710201
16	S 0710162	40		S 0710214
17	S 0710163	41		S 0710215
18	S 0710164	42		S 0710216
19	S 0710165	43		S 0710217
20	S 0710170	44		S 0710218
21	S 0710172	45		S 0710219
22	S 0710173	46		S 0710220
23	S 0710174	47		S 0710221
24	S 0710175	48		S 0710222
Kit: RubG		Lot no:# AB782		Operator ID: FM

Figure 5.5   Work sheet for Rubella IgG assay, showing kit controls, IQC sample and test sample numbers including S0710214. Note that kit name and batch number, operator, date of assay and identifying number of IQC sample have all been recorded

the rest of the morning working in the same room as the processor, so he is able to ensure that there are no problems with the machine during the 2 hours that the assay is running.

Step 8: Reading the results. At 12.30 p.m., Mr FM returns to the processor to collect the results, which have been uploaded from the machine to a linked computer. The computer programme calculates the ratio of positive to negative control readings and also flags any test readings which are of interest – in this case, readings which would be of interest are 'low positive' or 'negative'. Although he is still wearing a laboratory coat, he is not wearing gloves, in order to avoid risking contamination of the paperwork and the computer keyboard (Figure 5.6). He prints off the results (Figure 5.7) and first checks that the assay's internal controls give readings within acceptable limits. He then plots the value of the IQC on the chart (see Chapter 4) to make sure that the run is satisfactory from a quality point of view. As he is happy with all these parameters,

Figure 5.6 Specialist Biomedical Scientist using a computer to enter and check data relating to patients' results

he proceeds to record the results for each patient. The results are compared with the low positive control, which has a known concentration of anti-Rubella IgG of 10 IU/mL. This concentration is considered to represent a protective level of antibody within the patient.

5.3.3 Post-analytical phase

Step 9: Deciding whether further tests are required. Mr FM passes the printout of the results to his colleague, the senior Specialist Biomedical Scientist, Miss KS, telling her that there were no apparent problems with the run and that the IQC value passed Westgard's rules (see Chapter 4). Mr FM then goes for his lunch break, while Miss KS checks the results for the controls and agrees that they are satisfactory and then looks through the values for the test samples. Any results which are less than or equal to that recorded for the low positive control will need a follow-up test. The SOP states that these supplementary tests on negative and low positive samples must be carried out before sending out the reports on the whole batch. This is in case there has been a mistake such as in the order in which serum vials were placed in the analyser. Since there are potentially serious implications to being non-immune to Rubella, the biomedical scientists must be certain that they have issued the correct results

Results report: Lot: AB782				Rubella IgG Operator ID: FM			14/09/07
Location	Sample ID	Absorbance	Titre IU/mL	Location	Sample ID	Absorbance	Titre IU/mL
1	Blank	0.000		25	0176	2.999	> 10
2	NC	0.008		26	0177	3.117	> 10
3	NC	0.017		27	0178	2.876	> 10
4	NC	0.003		28	0190	3.011	> 10
5	HP	3.005		29	0191	2.589	> 10
6	HP	3.213		30	0192	2.765	> 10
7	HP	3.101		31	0193	3.201	> 10
8	LP	1.500		32	0194	3.204	> 10
9	LP	1.421		33	0195	2.909	> 10
10	LP	1.299		34	0196	2.856	> 10
11	IQC	2.901	> 10	35	0197	2.432	> 10
12	0148	3.032	> 10	36	0198	3.031	> 10
13	0149	2.345	> 10	37	0199	2.965	> 10
14	0150	2.970	> 10	38	0200	2.321	> 10
15	0151	2.789	> 10	39	0201	2.298	> 10
16	0162	2.634	> 10	40	0214	2.678	> 10
17	0163	2.590	> 10	41	0215	2.918	> 10
18	0164	3.178	> 10	42	0216	2.557	> 10
19	0165	2.876	> 10	43	0217	3.065	> 10
20	0170	3.146	> 10	44	0218	2.789	> 10
21	0172	2.453	> 10	45	0219	2.389	> 10
22	0173	2.657	> 10	46	0220	3.005	> 10
23	0174	2.199	> 10	47	0221	3.045	> 10
24	0175	2.908	> 10	48	0222	2.999	> 10

Figure 5.7   Printout of results showing controls and IQC

for each patient. If the follow-up tests give negative and low positive results for the same samples as the original assay, Miss KS can be confident that Mr FM had carried out his work correctly. In the case of today's anti-Rubella IgG EIA run, antibody at a higher concentration than the low positive control has been detected for every patient's sample (including sample number S0710214), so no further tests are indicated.

Step 10: Storing the specimen. When Mr FM returns after lunch, at 2 p.m., Miss KS tells him that she is happy with the results and that no further tests are required on those samples. He therefore files the serum vials in numerical order, in a tray at $-20\,°C$. These vials are kept for 12 months as they can prove useful if Mrs AP is exposed to any infection later during her pregnancy (for example, chicken pox). If protective levels of IgG against the pathogen of interest are found in the 'booking blood', then Mrs AP can be reassured that she is unlikely to contract the infection and her baby is not in danger. As for all patients' samples and tissues, this serum must be stored and eventually disposed of in accordance with the Human Tissue Act (see Box 5.5) Under the terms of this Act, a careful record of the location within the laboratory freezer of all these serum samples must be kept and they cannot be used for any research projects without ethical approval.

Box 5.5   The Human Tissue Act 2004

The Human Tissue Act 2004 was introduced to regulate the removal, storage and subsequent use of any human tissue (which is defined as material containing human cells). This Act covers activities in England, Wales and Northern Ireland, while the Human Tissue Act (Scotland) applies in Scotland.

The parts of the Act which affect diagnostic pathology departments can be summarized as:

1. The removal, storage and subsequent use of any samples taken from a person must be done only for strictly controlled and justifiable reasons ('scheduled purposes'), which includes diagnosis of disease.

2. Appropriate informed consent must be obtained for any use of the samples from either the person themselves, a relative (e.g. if material is being taken from a deceased person) or a parent or guardian (e.g. if material is being collected from a child).

3. If any work involving the stored tissue (e.g. a research project) which was not originally agreed at the time of collection is planned, the proposal must be scrutinized by an ethical committee of practitioners and lay people and further consent may be required.

4. Laboratories which store and use organs will be granted a licence and then monitored by the Human Tissue Authority (set up under the Act); they must meet set standards of practice and will be subject to inspection.

5. A licensed laboratory must designate a senior person (e.g. medical consultant, biomedical scientist manager) who will be held to account if there have been any breaches of the terms of the licence (e.g. someone has conducted unauthorized research) or any part of the Human Tissue Act has been broken.

6. When the agreed procedures have been carried out, stored material must be disposed of respectfully within a reasonable period.

Step 11: Validating the result. Miss KS uses the laboratory numbers to retrieve the original request forms from all the patients whose serum has just been tested for anti-Rubella antibodies. She carefully checks that the name and laboratory number which appear on the work sheet correspond to those on the request form in each case. At 3 p.m., when she is sure that everything is satisfactory, she uses a programme on the laboratory computer to validate the results, which means that it indicates that she judges them acceptable by pressing a certain key. Miss KS needs to read all the written information carefully. This means scrutinizing the scientific parameters of the assay and the patient details on the request form. Miss KS is Mrs AP's neighbour, so she has seen her anti-Rubella antibody results. However, as a biomedical scientist, Miss KS is bound by the Data Protection Act (see Box 5.6) and her codes of conduct (see Chapter 1) not to disclose the information she knows, but to leave it to the midwife, Miss JK, to discuss the results with Mrs AP, on her next visit to the ANC.

Box 5.6   Principles of confidentiality for data collected from patients

A.   Principles of the Data Protection Act 1998
The organization designated as the Data Controller (e.g. the NHS) must use data according to the following eight principles:

1. Processing of personal data must be done legally, fairly and because it is necessary to operate a comprehensive healthcare system.

2. Personal data must only be obtained for clearly specified reasons and not used for other purposes which are not covered by the law.

3. The personal data collected must be sufficient and appropriate to their purpose (and no more).

4. Personal data must be accurate and current.

5. Personal data must not be kept for longer than required to fulfil the purpose for which it was originally collected.

6. The legal rights (under the 1998 Act) of people from whom personal data are collected must be respected.

7. Systems must be in place to ensure that personal data cannot be accessed without authorization or used for unlawful purposes and that they cannot be accidentally lost or destroyed.

8. Personal data must not be transferred to a non-EU country unless safeguards against their misuse are in place.

B. Caldicott Principles

When disclosing information about a patient to another person, the healthcare professional should:

1. Make sure there is a good reason to do so.

2. Not identify the patient unless it is clearly necessary.

3. Only give the minimum amount of patient information.

4. Only share confidential information with people who really need to know it.

5. Make sure they know your responsibilities to protect patient confidentiality.

6. Make sure that they comply with the law (e.g. Data Protection Act 1998).

Step 12: Interpretation of the result. Clinical interpretation of test results is provided by medical consultants but, for many assays, they decide on a range of comments which senior Specialist Biomedical Scientists can use, and these are included in the results programme on the computer. In the case of routine anti-Rubella IgG results, the consultant microbiologist, Dr HL, is happy to allow senior Specialist Biomedical Scientists to use the approved comments on final results. In the situation where the antibody level is >10 IU/mL, the approved comment is 'Rubella IgG antibody detected. Immune to Rubella'.

Step 13: Authorizing and issuing the result. Once Miss KS has added the comments, she authorizes the results as the final version. They are now available on the computer to users outside pathology (for example, Miss JK, the midwife in the ANC) and the information from this batch is also sent for printing of paper copies on the printer in Mrs EM's office (see Figure 5.8). She collates them and puts them in envelopes according to the destination of the report. The results for the ANC, which include Mrs AP's result, are ready at 4.30 p.m. and Mr DH the porter takes them to the clinic on his first visit the following morning. Thus, the anti-Rubella antibody result will be filed in Mrs AP's notes, ready for her next visit to the ANC clinic a few weeks later.

AN Other NHS Trust

Microbiology Report

Patient Name: **Mrs Ann Pregnant**   DOB **14/08/82**
                                   Hospital Number **JKL 123456**
Ward **ANC**   Consultant code **DAC**
Specimen: **Blood – serum**   Date received: 12/09/07

Laboratory number: **S0710214**

Interim report
Result:

Rubella IgG antibody DETECTED >10 IU/mL
IMMUNE TO RUBELLA

Results awaiting confirmation: HepBsAg, Toxo IgG, TPHA, HIV

Figure 5.8   Result report form for Mrs AP

5.4 Health and safety in the clinical laboratory

It has long been recognized that it is a good idea to look after employees properly while they are work. This means considering both their mental and physical well-being. For example, a very stressed person will not work efficiently and may, paradoxically, take longer to complete specific tasks than someone who is not feeling too pressured. Similarly, if an employer decides not to spend money to replace faulty equipment, they could find that their employee has to take extended sick leave due to an injury. In the United Kingdom, the welfare of employees and visitors to a particular workplace is governed by law under the Health and Safety at Work (HSAW) Act (1974). This states that the employer has an obligation to provide all appropriate training, guidelines and protective equipment, and the employees have a responsibility to use them. The requirements of this law for employers and employees are summarized in Box 5.3. This law is supported by the work of the Health and Safety Commission, which was set up to give guidance on the HSAW Act, by writing and approving codes of conduct. Also, there is the Health and Safety Executive (HSE), which is there to enforce the Act and investigate cases where a breach of the Act is reported. Accounts in the news of major accidents in a workplace (for example, when a building collapses and injures people who were working inside it) will often report that the HSE have begun an investigation. The aim will be to investigate whether the HSAW Act has been breached by either the employer or employees and, if so, to call the offenders to account.

The clinical laboratory is a potentially hazardous environment and all biomedical scientists should behave professionally and sensibly at all times, in order to avoid putting themselves or others at risk of injury or infection. There are five main hazards to consider: biological, chemical, electrical, physical and environmental.

1. *Biological hazards.* All samples may contain infectious agents, although these are not necessarily the focus of the test procedure which has been requested. It is better to treat all samples as though they contain dangerous pathogens, wear personal protective equipment (i.e. laboratory coat, gloves and eye protection) when handling them, wash one's hands before leaving the laboratory and be careful at all times. It may be tempting not to bother about wearing gloves to prepare a routine blood slide in haematology, but this it is not good laboratory practice. The patient could be carrying a blood-borne infection which may or may not have been diagnosed. Some assays use biological reagents such as antibodies; these should have been prepared so that they are safe to use in the laboratory, according to the appropriate SOPs, but again it is better to err on the side of caution and take extra care. Some samples pose a known higher risk to staff handling them, such as respiratory samples from a patient with lower respiratory tract infection. The causative agent could be a number of organisms, including *Mycobacterium tuberculosis*, and therefore particular measures are taken when handling such samples. The Advisory Committee on Dangerous Pathogens (ACDP) publishes a list of microorganisms, allocated to 'categories' according to how infectious they are and how easy it would be to treat a laboratory-acquired infection. They advise on precautions which should be in place for the safe manipulation of organisms in each category. General procedures in the open laboratory are appropriate for category 2, whereas the respiratory sample in the example above must be handled in category 3 facilities. Hence the term widely used term 'cat. 3' to mean a potentially dangerous organism.

2. *Chemical hazards.* Most laboratory procedures involve the use of chemical reagents at some stage. Some of these chemicals, such as formalin, are dangerous poisons and swallowing a small amount, inhaling the vapour or, in some cases, allowing contact with the bare skin, may be rapidly fatal. Many of the stains used to allow the examination of changes in blood cells and tissues or identification of microorganisms are potential carcinogens if the person is exposed to them over a prolonged period, and others are skin irritants. Some chemical agents are only a danger to some people, and the latex which is used to make gloves is a good example of that. Although some people develop anaphylactic shock on exposure to latex particles, it is rare. However, it is common to develop contact dermatitis

when regularly using latex gloves coated with powder. This is estimated to affect 15–20% of healthcare workers, including biomedical scientists, which is why vinyl gloves are widely used and should always be available to staff in pathology laboratories.

3. *Electrical hazards.* Laboratories contain a wide range of electrical equipment, from small refrigerators to large analysers. Although some problems with a machine are easy to detect (e.g. it stops working!), a fault in the wiring may be less obvious. Some electrical defects can be very dangerous and could lead to someone being electrocuted. In order to prevent this, even the simplest piece of electrical equipment brought into the workplace (for example, a domestic kettle) should be checked by a qualified electrician before being used. When this has been done, a label should be attached to the equipment stating the date of the electrical test and setting a date for re-test (for example, 12 months later.)

4. *Physical hazards.* The laboratory can be a confined space, containing large analysers, in between long benches, which are surrounded by cupboards, refrigerators, freezers, incubators and computers. When designing or re-arranging the layout of a laboratory (for example, to accommodate a new machine), it is very important to ensure that there is plenty of space for people to move around safely within each room. Section Leaders should be aware of potential dangers, but it is also important for all staff to be sensible and careful not to create physical hazards for their colleagues. For example, they should not leave boxes in the middle of a room for others to fall over or allow electrical wires to trail across the floor, and they should make sure that they close cupboard doors.

5. *Environmental hazards.* Since clinical laboratories handle infectious agents and toxic chemicals, it is usually not possible to open the windows to let in fresh air. This means that ventilation relies on the circulation of air through an air conditioning system. This must be monitored and regularly maintained to ensure that the air quality is good; the laboratory should not be too cold or too hot for people to work in and the air conditioning system itself should not pose an infectious hazard (for example, from Legionnaire's disease). Some pathology departments are in bright, airy, modern buildings which allow in plenty of natural light, whereas others are in dark corners of old hospitals. In addition to being depressing places to work in, old buildings sometimes contain high levels of radon in their walls, so this should be checked. It is becoming less common in diagnostic laboratories to use assays which involve radioactive reagents; where they are used, they must be stored safely and all who perform the tests must wear a badge containing X-ray film on their laboratory coats which is regularly monitored to ensure that they have not been exposed to excessively

high levels of radioactivity. Another more mundane environmental hazard is a gas leak, which is particularly likely to happen in a laboratory where Bunsen burners are used.

Any accidents, incidents which might have led to an accident or laboratory-acquired infections that occur in the laboratory must be recorded within the department and reported centrally within the employing organization (e.g. the NHS Trust). The employer has a legal obligation to report fatalities, major incidents, any accidents where a person has serious injury requiring more than 3 days off work, outbreaks of diseases related to work, incidents involving gas and if something dangerous happens, to the HSE under the Reporting of Injuries, Diseases and Dangerous Occurrences Regulations 1995 (RIDDOR). This is to allow data to be compiled on accidents and incidents for each employer and at a national level, which may point to trends in dangerous practices and perhaps indicate that there is a need for new health and safety guidance in a certain area of work.

5.5 Confidentiality, the Data Protection Act 1998 and the Caldicott Principles

The meaning of 'confidentiality' in a professional context is keeping a piece of information private. In today's society, we all have to provide a lot of information about ourselves which we might not want to be made public, in order to buy goods and services. Much of this information is held on electronic databases and is available to a lot of individuals who we never actually meet (for example, people working on banking enquiries at a call centre). Confidentiality is therefore extremely important to protect vulnerable people and to ensure that the professionals with access to the information do not exploit it for personal or financial gain. The role of the Data Protection Act 1998 is to provide this protection and to uphold the principle of confidentiality. All healthcare workers have access to information which the patient would not want to be public knowledge – from their date of birth to their diagnosis.

Under the terms of the Data Protection Act 1998, confidential information which patients provide during the course of their interactions with health services fall into two categories:

- *Personal data*: these are defined as information about a *living person* by which they could be identified or data which could be put with other data that are already available which could lead to identification of an individual such as someone's name, address, age, gender, race and nationality.

- *Sensitive personal data*: these are defined as information about someone's ethnic origin, religion or other beliefs, sexuality, health (physical or mental), membership of trade unions, political opinions, legal summons or convictions.

All such data which are given by patients and recorded either on paper or electronically and are processed in any way (which includes being stored on file) are subject to the terms of the law. There is a Data Protection Commissioner whose job is to make sure that the data protection system is not abused and who makes the final decision about complaints by members of the public. Each organization which collects and uses data covered by the Act (e.g. the NHS) is called a data controller and it must make sure that all its employees are aware of and keep to the law according to the terms of the eight principles given in Box 5.6. Except in certain circumstances, the individuals on whom personal data is stored have the right to access the information about themselves wherever and however it is stored. The exception might be a person who was seriously mentally ill and for whom the knowledge that some information was being stored about them might be detrimental to their condition. Although a person has the right to see the information which is kept about them by a particular data controller, they are required to make an appointment to see it and to pay a fee. This is because someone has to look for the records, which might be stored away from the main area of work or on a computer which would need to be made accessible. Also. in some cases, for example looking at medical notes, a professional needs to be available to answer any questions which the viewer might have (e.g. about medical terminology).

In 1997, a review of the flow of information about patients within the NHS was set up and the resulting Caldicott Report recommended that clear guidance and protocols should be in place about data which are passed between professionals within the NHS and from the NHS to other organizations (e.g. those involved in social care). This is so that only information which is strictly necessary is passed on and everyone who has access to data and transmits it knows their limits and responsibilities. They recommended that a senior person within each organization (e.g. NHS Trust) should be appointed the 'Caldicott Guardian' and should be should be responsible for ensuring that patient confidentiality is secure, through training programmes for staff and robust procedures; this idea has been widely taken up. The committee's proposal that the patient's NHS number (which is unique and constant and would therefore be the same in each hospital they attended) should become the main way of identifying them has not. The outcome of the report was a list of Caldicott Principles which NHS staff should abide by when processing patient data, and these are also given in Box 5.6. Guidance on confidentiality for NHS staff is also provided regularly; up-to-date information can be found on the relevant Department of Health website (see reference list). Biomedical scientists must be aware of their responsibility to keep the information to which they have access confidential. They should only discuss results from a particular patient with colleagues involved in the diagnosis. This does not mean that they cannot talk about interesting findings or events during the day outside work, but that they must do it in a general way, so that individual patients cannot be identified. A good way of maintaining high professional standards in the area of confidentiality is to consider how you would feel if you were the patient whose data were being shared.

Quick quiz

1. Why is all work in the clinical laboratory guided by written protocols?

2. What does the acronym SOP stand for?

3. Name the three stages of the progress of a clinical specimen through a laboratory.

4. Give a definition of the 'minimum data set' for patient identification.

5. What is the minimum data set for patient identification for a sample sent for cross-matching? Why are the checks on this type of sample more rigorous than usual?

6. Why is it best to provide clear plastic bags with two pockets to transport samples from the ward to the laboratory?

7. What is PPE?

8. What is the purpose of the laboratory number assigned to each sample?

9. Which staff group is authorized to provide clinical comment on test results, and why?

10. What obligations do employees have under the Health and Safety at Work Act 1974?

11. Describe the role of the Health and Safety Executive (HSE).

12. Name the five main hazards which staff may encounter in a clinical laboratory.

13. What does the acronym COSHH stand for?

14. Name two kinds of incident or accident which must be reported to the HSE under the Reporting of Injuries, Diseases and Dangerous Occurrences Regulations 1995.

15. List three pieces of information which are classed as 'personal data' under the Data Protection Act 1998.

16. What is 'sensitive personal data'?

17. Who or what is a Data Controller?

18. Give three of the Caldicott Principles.

19. Why is it important for a biomedical scientist to inform their manager immediately if they are unable to attend the laboratory for work?

20. What are the two main reasons why a biomedical scientist should not disclose test results to a patient even if they are relative or close friend?

Suggested exercises

1. Perform a health and safety audit by choosing one section of the laboratory that you are working in where your colleagues give permission for you to watch them for 1 hour. Find a safe position and make a detailed observation of everything that happens in that period, for example by taking notes or making a video. Then review all events in the light of what you have learned about health and safety, noting clear examples of good practice, obvious poor practice and anything which you observed which you think might be unsafe but are unsure about. Discuss your findings with a senior colleague.

2. Compile a list of all the information that you have access to within the laboratory in your current position as a student or trainee biomedical scientist. For each type of information, indicate whether it is confidential or not, giving your reasons. Then compile a second list of the extra information that you would have as a registered practitioner, again indicating with reasons whether or not it should be treated as confidential.

3. Select a SOP for an assay which is performed regularly in the laboratory that you are working in and make a copy of it. Read through it thoroughly and think about why each piece of information has been included. Annotate your copy to indicate your thoughts about this SOP, referring to any relevant laws (e.g. HSAW) or policies (e.g. Trust incident reporting procedure) at each stage.

Suggested references

Advisory Committee on Dangerous Pathogens (2004). *The Approved List of Biological Agents 2004*. Norwich: HMSO; www.hse.gov.uk/pubns (this gives the lists of organisms in each containment category).

Data Protection Act 1998: www.opsi.gov.uk/ACTS/acts1998.
Departments of Health for the United Kingdom websites:

England: www.dh.gov.uk/en
Wales: www.wales.nhs.uk
Scotland: www.show.scot.nhs.uk
Northern Ireland: www.dhsspsni.gov.uk

Health and Safety Executive website: www.hse.gov.uk. This contains information on:

- The Health and Safety at Work Act 1974: www.hse.gov.uk/hsaw.

- Control of Substances Hazardous to Health Regulations 2002: www.hse.gov.uk/coshh.

- Reporting of Injuries, Diseases and Dangerous Occurrences Regulations 1995: www.hse.gov.uk/riddor.

Human Tissue Authority website: www.hta.gov.uk.
Human Tissue Act 2004: www.hta.gov.uk/human_tissue_act.
Institute of Biomedical Science (2000). *Giving Results Over the Telephone*. London: IBMS; www.ibms.org.
Institute of Biomedical Science (2003). *Patient Sample and Request Form Identification Criteria*. London: IBMS; www.ibms.org.
Serious Hazards of Transfusion (SHOT); www.shotuk.org.

6

Introduction to the biomedical science disciplines

6.1 Introduction

The work in clinical pathology departments encompasses a wide range of techniques, across a number of specialist areas. Due to the depth and breadth of knowledge and expertise required, biomedical scientists (and their clinical scientist and medical colleagues) tend to train and work in one discipline. As mentioned in Chapter 2, the main disciplines are clinical biochemistry, haematology and transfusion science, histopathology and cytopathology and medical microbiology. The range of tests offered within these laboratories provides the necessary support for diagnosis and treatment in most general hospitals. However, there are also a number of specialist areas on which biomedical scientists can concentrate, such as immunology, virology and genetics. In some hospitals, there will be separate laboratories offering a service specifically in these specialities, according to size and circumstances. For example, a hospital which carries out a large number of transplants or provides regional expertise in allergy testing would have a separate immunology laboratory where the required range of assays will be available; in a smaller hospital, some simpler immunological tests would be carried out in the other departments, whereas samples requiring more complex immunological assays would be sent to a reference laboratory. Similarly, some medical microbiology departments carry out basic virological tests, while other hospitals have distinct virology departments. Figure 6.1 illustrates how the specialist areas and the main disciplines relate to each other.

The way in which the pathology service is arranged can also vary, particularly as services are rationalized (see Chapter 2) and as technology changes. To take full advantage of these changes, biomedical scientists must be multidisciplinary in their outlook. Large and versatile analysers are increasingly available, which can test for a wide range of analytes in blood, covering several of the traditional discipline boundaries (e.g. clinical biochemistry, haematology and microbiological serology). In some hospitals, amalgamated departments of 'Blood Sciences'

An Introduction to Biomedical Science in Professional and Clinical Practice   Sarah J. Pitt and James M. Cunningham
© 2009 John Wiley & Sons, Ltd

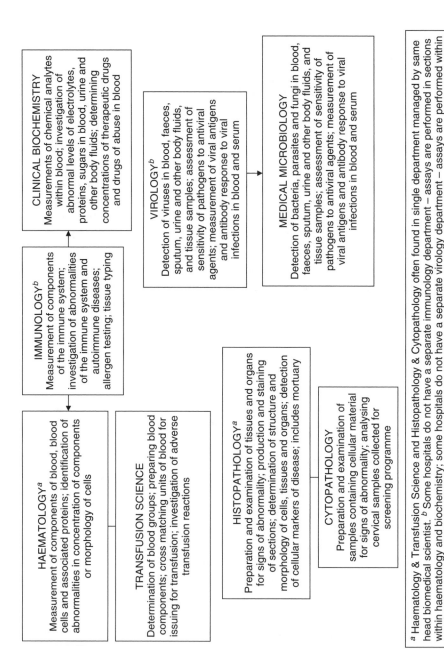

CLINICAL BIOCHEMISTRY
Measurements of chemical analytes within blood; investigation of abnormal levels of electrolytes, proteins, sugars in blood, urine and other body fluids; determining concentrations of therapeutic drugs and drugs of abuse in blood

IMMUNOLOGY[b]
Measurement of components of the immune system; investigation of abnormalities of the immune system and autoimmune diseases; allergen testing; tissue typing

VIROLOGY[b]
Detection of viruses in blood, faeces, sputum, urine and other body fluids, and tissue samples; assessment of sensitivity of pathogens to antiviral agents; measurement of viral antigens and antibody response to viral infections in blood and serum

MEDICAL MICROBIOLOGY
Detection of bacteria, parasites and fungi in blood, faeces, sputum, urine and other body fluids, and tissue samples; assessment of sensitivity of pathogens to antiviral agents; measurement of viral antigens and antibody response to viral infections in blood and serum

HAEMATOLOGY[a]
Measurement of components of blood, blood cells and associated proteins; identification of abnormalities in concentration of components or morphology of cells

TRANSFUSION SCIENCE
Determination of blood groups; preparing blood components; cross matching units of blood for issuing for transfusion; investigation of adverse transfusion reactions

HISTOPATHOLOGY[a]
Preparation and examination of tissues and organs for signs of abnormality; production and staining of sections; determination of structure and morphology of cells, tissues and organs; detection of cellular markers of disease; includes mortuary

CYTOPATHOLOGY
Preparation and examination of samples containing cellular material for signs of abnormality; analysing cervical samples collected for screening programme

[a] Haematology & Transfusion Science and Histopathology & Cytopathology often found in single department managed by same head biomedical scientist. [b] Some hospitals do not have a separate immunology department – assays are performed in sections within haematology and biochemistry; some hospitals do not have a separate virology department – assays are performed within microbiology

Figure 6.1 Organization of discipline specific laboratories within a typical hospital pathology department

are being created, where biomedical scientists from all the individual disciplines work together as a larger team and are required to learn about tests and analytes from the other areas, with which they might be less familiar. Advances in molecular biology techniques also make this a realistic option for diagnostic tests based on the detection and analysis of nucleic acids; once the machinery and assay protocols are operating reliably, they can be adapted to detect a range of molecular markers, from the genetic material of microorganisms to abnormal gene sequences in a patient's cells. The trend towards offering more patient-centred 'point of care testing' (see Chapter 2) opens the possibility for biomedical scientists to take up extended roles in primary care, but again this will require a working knowledge of routine tests conventionally undertaken in separate laboratories within a hospital pathology service.

Current undergraduate biomedical sciences programmes are designed to be multidisciplinary, so recent graduates should have some knowledge about all aspects of pathology. However, placements are often organized so that students and trainees tend to acquire the majority of their clinical laboratory experience in a single discipline. This is acceptable, since the skills required to register as a practitioner with the Health Professions Council are generic and evidence of competency (see Chapter 8) can be acquired in any clinical pathology laboratory. However, in order to operate effectively in the multidisciplinary pathology team and to respond to future changes in the management of clinical laboratory services, it is important for biomedical scientists to have an appreciation of the work that is carried out in all laboratories.

This chapter will provide an overview of the work of each discipline, by discussing examples of diagnostic, monitoring and screening assays from each. It will provide the reader with some insight into the work of the laboratories in which they have no experience, but does not provide scientific or technical detail, which can be found in other books (see Suggested references).

6.2 Haematology and transfusion science

This is essentially the study of blood, its component proteins and blood cells. Biomedical scientists working in haematology and transfusion science do this by observing the reactions in samples of whole blood or plasma when they are mixed with certain reagents. They also make blood films in order to examine the cellular components of blood. Blood films are stained with a Romanowsky-based stain (such as Giemsa, Leishman's or May Grunwald), which stains nuclear material purple and cytoplasm blue, so that cells can be seen and identified. Abnormalities in red cell morphology or relative proportions of certain white cells can thus be found in a patient's blood, as well as parasites such as *Plasmodium* spp. Box 6.1 lists some of the common haematology and transfusion science tests and their reference ranges, where appropriate.

Box 6.1 Examples of some commonly performed haematology and transfusion science tests

	Reference range (if applicable)
Full blood count:	
Haemoglobin concentration	11.5–16.5 g/dL for adult female
	13.5–18.0 g/dL for adult male
Erythrocyte count	$3.8–5.8 \times 10^{12}$/L for adult female
	$4.5–6.5 \times 10^{12}$/L for adult male
Leucocyte count	$4.0–11.0 \times 10^{9}$/L (adult)
Erythrocyte sedimentation rate	0–15 mm/hour
Prothrombin time	12–15 seconds
International normalized ratio	0.8–1.2 (normal adult)
	INR 2.0–4.0 (patient on anticoagulation therapy)
Activated partial thromboplastin time	27–32 seconds
Activated partial thromboplastin time ratio	0.8–1.2
Platelet count	$150–450 \times 10^{9}$/L (adult)
Blood grouping	

6.2.1 Diagnosis – full blood count (FBC)

The full blood count (FBC) is a test which is commonly performed to help with diagnosis. For example, in a situation where the patient is lethargic, this might be due to anaemia, so it is useful to establish the haemoglobin level; further examination can help to ascertain the underlying cause (for example, it might be iron deficiency or an abnormality in the morphology of the red blood cells). The FBC test involves determining the concentration of haemoglobin, red and white blood cells, reticulocytes and platelets in a sample of a patient's blood. Although these parameters can all be assayed separately using manual tests (such as using a counting chamber to calculate red and white cell levels – see Chapter 7), it is usual practice to examine samples in a single automated analyser (see Figure 6.2), due to the large volume of requests for FBC per day.

To determine haemoglobin (Hb) concentration, an aliquot of the patient's blood is mixed with a chemical (usually cyanmethaemoglobin or sodium lauryl sulphate) inside the analyser, the absorbance is read at an appropriate wavelength and the machine uses this to calculate the Hb level. Most analysers use

(a)

(b)

Figure 6.2 An automated cell count analyser. (a) Racks of blood tubes being loaded on to the analyser for automatic sampling; (b) a biomedical scientist presents a blood tube to the analyser for manual sampling

either 'aperture impedance' (Coulter method) or 'light scatter' or a combination of both to quantify the concentrations of erythrocytes (red blood cells), leucocytes (white blood cells), reticulocytes and thrombocytes (platelets) in blood samples. In aperture impedance, a known volume of the blood is forced through a narrow tube within the machine which has a hole ('aperture') at the end. An electric current is passed across this aperture; when a cell crosses its path, it interferes with ('impedes') the flow of the current, which can be detected by a change in voltage. The size of the cell affects the strength of the voltage change and the concentration of each cell type is calculated by the number of cells detected in that known volume. The light scatter technique is similar, expect that a laser beam is shone across the aperture instead of an electric current; the amount and direction of light scattered depend on the size and shape of the cell. Figure 6.2 shows a typical automated cell count analyser.

Readings outside the normal range are flagged by the machine and may need further investigation. For example, if the patient's specimen appeared to have an abnormally high white cell count, the biomedical scientist would prepare and stain a thin blood film to do a manual differential count (see Figure 6.3).

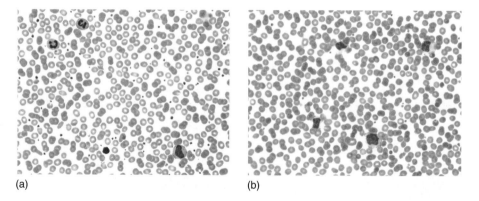

(a)   (b)

Figure 6.3 (a) Thin blood film from normal patient showing two neutrophils, one lymphocyte and one monocyte; (b) thin blood film from a patient with leucocytosis, showing abnormally high levels of atypical lymphocytes. Both photographs courtesy of Mrs S. Bastow

6.2.2 Monitoring – coagulation assays: prothrombin time (PT) and activated partial thromboplastin time (APTT)

In the formation of blood clots, to prevent bleeding, platelets are activated and a series of biochemical reactions involving proteins called 'clotting factors' is set up, the end result being a fibrin clot. There are two possible 'pathways' through this 'coagulation cascade', called the intrinsic and extrinsic pathways, depending on the initial trigger. Some patients are unable to form fibrin clots to prevent excessive bleeding, due to an acquired or inherited deficiency in one of the clotting factors. For example, in haemophilia A, an inherited genetic mutation means that the person is unable to make Factor VIII, which is a component of the intrinsic pathway.

Clots can sometimes develop internally and these may travel around in peripheral blood until they cause a blockage in important blood vessels. For example, a deep vein thrombosis (DVT) in the leg is a complication of inactivity, such as being confined to bed due to illness or sitting in a confined space on a long-haul aeroplane flight. Patients with DVT are treated with an anticoagulant (such as heparin or warfarin), which works to increase clotting times; they must be carefully monitored while on this treatment, in case the dose they are taking is providing too little or too much anticoagulant activity – both of which could be dangerous.

Biomedical scientists contribute to this monitoring by measuring the prothrombin time (PT), which assesses the extrinsic pathway, and the activated partial thromboplastin time (APTT), which assesses the intrinsic pathway, in a sample of patient's plasma. The tests are performed by mixing an aliquot of plasma with a reagent containing calcium ions and either thromboplastin (PT) or phospholipid (APTT) in a test-tube and noting the time it takes for the formation of a visible clot. Although this can be done by eye, using a stopwatch, processing a large number of samples at a time would be difficult and so automated machines have been developed which can do the required assays, such as that shown in Figure 6.4. Some analysers 'read' the clot by shining a beam

(a)

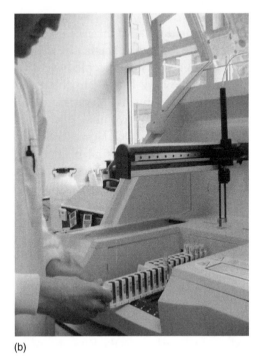

(b)

Figure 6.4 (a) An automated coagulation analyser; (b) a biomedical scientist loads samples into the analyser

of light on the test well and detecting when and how light from this beam is scattered; other models detect the extent to which light is absorbed.

The PT is used to monitor patients taking anticoagulants, but the results can be inconsistent, due to variability in the test reagents. Therefore, haematology departments use a method of standardizing readings, called the International Normalized Ratio (INR), and give results in this format. The equation used to calculate the INR from the PT reading is

$$INR = \frac{PT}{MNPT} \, (ISI)$$

where PT = prothrombin reading, MNPT = mean normal prothrombin time for the local population and ISI = international sensitivity index for the thromboplastin reagent being used.

Using the PT from a patient with a normal clotting time, the INR calculation should give a ratio of 1. Patients on anticoagulant therapy need to keep their INR between 2.0 and 4.0. If the test results indicate a ratio outside these limits, then they must adjust their dose of heparin or warfarin accordingly.

6.2.3 Screening – determining blood group

The main blood grouping system is called the ABO system. It relies on the fact that humans inherit the A or B antigens genetically and these are expressed on the surface of their red blood cells. Some people have A antigens on their red cells (blood group A) and some have B antigens (blood group B). People can also have both A and B red cell antigens (blood group AB) or they can have neither (blood group O). The gene has two alleles, so an individual's genotype will be either AA, AB, AO, BB, BO or OO. However, the O blood type is coded for by a recessive gene, so there are only four phenotypes, namely A, B, AB and O. People who express either A or B red cell surface antigens also make antibodies to the opposite antigen – so a person with blood group A has anti-B antibodies circulating in their peripheral blood. However, a person with blood group AB has neither anti-A nor anti-B antibodies under normal circumstances. Someone with blood group O produces both anti-A and anti-B antibodies. The other important blood group system is the Rhesus (Rh) system, which based on the presence or absence of the D antigen on red blood cells; most people are Rh positive. It can be very dangerous to transfuse donated blood from one person to another without first ascertaining whether their blood groups are compatible (see Table 6.1). It is acceptable to give blood from a person with blood group O to a person with blood group A, as long as the Rh status of donor and recipient is compatible; however, the reverse donation is not possible, as the recipient would make anti-A antibodies in response to the presence of the 'foreign' A antigen, which would destroy the donated red blood cells. Similarly, when blood from an Rh negative person is mixed with Rh positive blood, they develop anti-D

Table 6.1 ABO and Rh blood groups and acceptable donor and recipient combinations

Blood group		Can donate to groups:	Can receive from groups:
A	Rh+	A Rh+, AB Rh+	A Rh + /−, O Rh + /−
B	Rh+	B Rh+, AB Rh+	B Rh + /−, O Rh + /−
AB	Rh+	AB Rh+	A Rh + /−, B Rh + /−, AB Rh + /−,O Rh + /−
O	Rh+	A Rh+, B Rh+, AB Rh+ O Rh+	O Rh + /−
A	Rh−	A Rh + /−, AB Rh + / − [a]	A Rh−, O Rh−
B	Rh−	B Rh + /−, AB Rh + /−	B Rh−, O Rh−
AB	Rh−	AB Rh + /−	A Rh−, B Rh−, AB Rh−,O Rh−
O	Rh−	A Rh + /−, B Rh + /−, AB Rh + /− O Rh + /−	O Rh−

[a]Rh +/− indicates that either Rh positive or Rh negative blood would be acceptable.

antibodies. This is particularly hazardous when the mixing of blood occurs in the umbilical cord during pregnancy where the mother is Rh negative, but the baby is Rh positive. The current baby is usually unaffected, but during any subsequent pregnancies, the maternal anti-D will have an adverse effect on foetal blood cells.

There are a number of other red cell antigens and antibodies which affect the outcome of a blood transfusion, but the initial screening to determine someone's blood group involves the ABO and Rh systems. Four separate aliquots from the person's blood sample are mixed with either anti-A antibody, anti- B antibody, anti-D or anti-D2 (a common variant of D). Thus, if the original sample contains red blood cells expressing A antigen, they will react with the anti-A antibodies (but not anti-B), causing visible agglutination; this will indicate that the person's blood group is A. These reactions can be undertaken in a series of test tubes or 96-well microtitre plates, but a test method using a gel matrix is often preferred, as it is relatively rapid and reduces the risk to operators from blood-borne infections. The test format is a series of enclosed, but transparent, tubes attached to a card (see Figure 6.5) which are filled with the gel. They are labelled A, B or D, according to which (monoclonal) antibody has previously been added to the gel. An aliquot of the patient's blood is added to each tube and the card is centrifuged, which pulls the blood cells through the matrix. If the patient's blood cells meet the corresponding antibody within the gel, they will react together and become stuck with the matrix, producing a line which can be recorded. If there is no antigen – antibody reaction, then the patient's cells fall to the base of the tube and are seen as a solid plug (as shown in Figure 6.5). The procedure can be automated to allow processing of a number of patient samples simultaneously.

The result is then usually checked by mixing aliquots of the person's serum (or plasma) with one each of the following: group A red cells expressing A

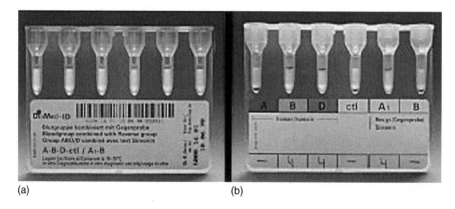

(a)   (b)

Figure 6.5 Gel agglutination system for determining ABO and Rh blood group. (a) Arrangement of tubes within the card before addition of patient sample; (b) result from a patient with blood group B Rh+. Picture courtesy of Mrs A. Trezise

antigen, group B red cells expressing B antigen or group O red cells expressing no antigen. A person whose blood group is A should have anti-B antibodies in their serum, and again this reaction will lead to agglutination.

6.3 Clinical biochemistry

This is concerned with the analysis of chemicals within the body and establishing whether any are at unusually high or low levels. Chemical analytes found in a patient's specimens may occur naturally, as part of the normal metabolism, such as sodium or potassium, or may have been ingested by the patient deliberately (e.g. prescribed medication, illegal drugs) or by accident (e.g. unwitting exposure to toxic chemicals). Biomedical scientists in biochemistry (clinical chemistry) laboratories measure the level of chemicals in patients' blood (through assays on serum or plasma), urine and occasionally other body fluids. Some examples of common biochemical analytes and reference ranges are shown in Box 6.2.

Box 6.2 Examples of some analytes which are commonly assayed in clinical biochemistry laboratories

	Reference ranges
Urea and electrolytes in plasma (adults)	
Urea	1.7–8.3 mmol/L
Sodium	135–146 mmol/L

Potassium	3.2–5.1 mmol/L
Creatinine	44–80 µmol/L (in females)
	62–106 µm/L (in males)
Liver function tests	
Bilirubin	3–21 µmol/L (adults)
Total protein	60–82 g/L
Albumin	30–45 g/L
Alkaline phosphatase	20–90 U/L (adults)
Alanine aminotransferase	10–40 IU/L
γ-Glutamyl transferase	5–35 IU/L (females)
	10–55 IU/L (males)

6.3.1 Diagnosis – multiple myeloma

Myeloma is a malignancy which causes overproduction of plasma cells in the bone marrow. Plasma cells make and secrete antibodies and in this disease unusually high concentrations can be found in the patient's blood and urine. Since each plasma cell only produces a single clone of antibody, then these excessive antibodies will be monoclonal; they are called 'paraproteins'. The presenting symptoms for multiple myeloma are often fairly non-specific (including back pain, lethargy and increased susceptibility to infectious diseases), so laboratory tests make an important contribution to diagnosis.

Initial tests carried out by biomedical scientists on the patient's serum would show raised levels of total protein and increased levels of IgG antibody. The individual components of the serum protein must then be separated by agarose gel electrophoresis. This can be done according to the combination of amino acids making up each type of protein, which will give the molecule an overall electric charge. When the proteins are placed in a matrix through which they can migrate, in a suitable ionizing buffer, which is subjected to electric current, they will move according to the net electric charge and size of the molecule. Serum proteins are usually separated using agarose gel as the matrix. This is poured into a transparent rectangular-shaped box, leaving slots to add the serum samples at one end and covered with buffer to make sure the process takes place at the correct pH (see Figure 6.6a). Patients' and control sera are mixed with a dye and then aliquots are placed into the slots in the gel. An electric current is passed through the gel, by placing a positive electrode at one end and a negative electrode at the other. The position that each type of protein has reached after a specified amount of time is evident from the location of bands

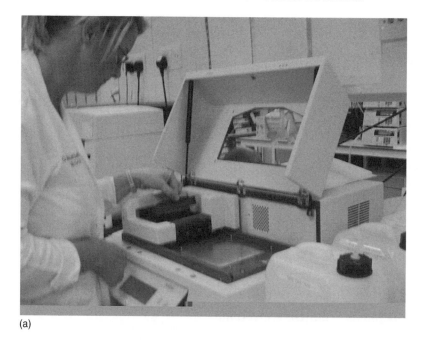

(a)

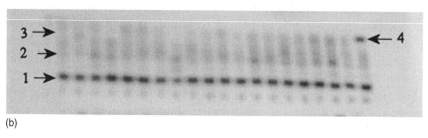

(b)

Figure 6.6 (a) A biomedical scientist loads the comb into an agarose gel within the electrophoresis apparatus. (b) Result from serum protein electrophoresis showing control serum on the left-hand side and serum from patient with myeloma on the right-hand side. Arrows indicate position of serum proteins: 1, albumin; 2, α_1-, α_2-, β_1- and β_2-microglobulin; 3, γ-globulin; 4, abnormally high concentration of γ-globulin in serum from a patient with myeloma

of the dye within the gel. The patterns obtained for patients' serum samples can be compared with the control sample, which contains known concentrations of certain proteins that are expected to migrate in a particular order. In myeloma, paraprotein will be detected in the patient's serum (see Figure 6.6b).

Further tests can be then be performed to identify the exact nature of the paraprotein (e.g. by scanning densitometry of the sample and immunological assays on the electrophoretic band of interest). Electrophoresis would also be carried out on a sample of the patient's urine to test for the presence of abnormal antibodies (Bence Jones protein).

6.3.2 Monitoring – thyroid function tests

Disorders of thyroid function are reasonably common in middle age and occur more frequently in women than men. They can happen unexpectedly, with no apparent cause ('idiopathic'), or as result of treatment for another medical problem ('iatrogenic'). Thyroid dysfunction can be manifest as hyperthyroidism (i.e. the patient has an overactive thyroid) or hypothyroidism (i.e. the patient's thyroid is underactive).

There are a range of possible treatments, from waiting to see whether the condition will spontaneously reverse through to removal of the thyroid gland altogether. Many patients with poorly functioning thyroid will be prescribed the hormonal drug thyroxine to stabilize their condition. It is important that such patients are monitored closely to make sure that their management and treatment are effective. This is achieved by measuring the levels of thyroid-stimulating hormone (TSH) in their blood.

The assay for TSH is a sandwich-type enzyme immunoassay (EIA). The solid phase is coated with antibody raised against TSH, to which patient's serum is added. The reaction well is incubated at 37 °C, to allow any TSH present in the serum to bind with the anti-TSH antibody. The well is washed and any reaction is then detected by adding anti-TSH antibody which has been conjugated to an enzyme. After a suitable period of incubation and washing, the substrate for the enzyme is added. This causes a detectable effect (e.g. colour change), which can be quantified and translated into the concentration of TSH present in the patient's serum (see Figure 6.7). This assay is usually performed in an

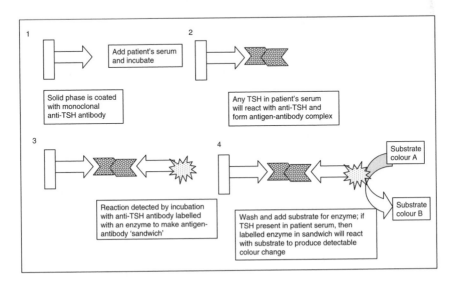

Figure 6.7   Principle of a 'sandwich' enzyme immunoassay test for thyroid-stimulating hormone in serum. In some automated analysers, whole blood is loaded on to the machine

automated analyser, so the biomedical scientist's role would be to load the sample into the machine and assess the results. If the result was outside the reference range, the clinician who requested the test would be informed so that they can review the patient's medication.

6.3.3 Screening – random blood sugar

There are two main types of diabetes mellitus, type I and type II. In both types, the patient has a deficiency of insulin and is unable to control their glucose metabolism adequately. Both 'hypoglycaemia' (markedly low blood sugar concentration) and 'hyperglycaemia' (excessively high blood sugar concentration) can be dangerous and potentially fatal. Although it is not fully clear why some people develop diabetes, both type I and II forms are thought to occur due to a complex interaction of genetic susceptibility and environmental factors. Type I is sometimes called 'juvenile onset' diabetes, because the condition is most often first recognized during childhood. It is a severe condition, resulting from the autoimmune destruction of the insulin-producing beta cells in the pancreas. The treatment is usually a daily injection of a preparation of the hormone insulin to correct the endocrine imbalance, coupled with careful management of the diet. This is to make sure that the insulin which has been injected into the blood stream has the correct amount of calories to deal with – too few and hypoglycaemia could occur and too many might allow hyperglycaemia to develop. Type II diabetes initially arises from a diminished sensitivity of cells and tissues to insulin and tends to become noticed in middle age; it is usually a less dangerous condition and can be effectively treated with drugs and sometimes simply a stricter control of the diet alone. Because it is uncommon for people with untreated type II diabetes to be acutely ill and go into a coma, many individuals do not notice or actually ignore their symptoms. Although not rapidly fatal, uncontrolled diabetes can be harmful in the long term, causing eye disease ('retinopathy'), degeneration of the nervous supply, particularly to the leg muscles ('neuropathy'), and problems with the circulation to the extremities causing loss of function and even gangrene in the toes. It is therefore important to identify people who might have, or be at risk of developing, type II diabetes and this can be achieved by testing their blood sugar levels as part of a general health screen. This is called a 'random blood sugar' test as it can be taken at any time and without particular preparation (i.e. eating or not eating for a specific length of time).

The test for random blood glucose is based on the activity of the enzyme glucose oxidase, which catalyses the reaction by which glucose is converted to gluconic acid and hydrogen peroxide (H_2O_2). This H_2O_2 is detected by a further series of reactions which leads to a colour change (see Figure 6.8).

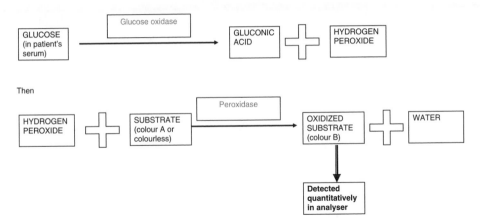

Figure 6.8   Chemical reactions in the glucose oxidase colorimetry method for quantifying blood glucose concentration

The test specimen is plasma, from blood which has been collected in a tube coated with fluoride – oxalate (to prevent any breakdown of the glucose before testing). The test is usually automated. The biomedical scientist would load the serum samples into the testing wells, along with the internal quality control serum (see Chapter 4) and a series of calibrators of known glucose concentration, from which the calibration curve can be constructed. The biomedical scientist will read and analyse the results and any samples which give a result outside the expected range (3.3–7.4 mmol/L for adults) will be reported promptly to the practitioner who conducted the health check. They will advise the patient to have further tests (e.g. fasting blood glucose) and thus decide on the best treatment and management of their condition.

6.4 Histopathology and cytopathology

This is the study of tissues and cells to detect any abnormal arrangements or morphologies which would indicate disease (pathology). The work of biomedical scientists in these disciplines involves preparing slides from pieces of tissue or specimens containing cells and then staining them. Samples received in the histopathology and cytopathology laboratory range from large organs which have been removed from bodies during *post mortem* examinations, to small pieces of tissue taken in a biopsy, to swabs taken to investigate cells in a certain areas of the body (such as cervical swabs). Biomedical scientists work closely with their medical histopathologist colleagues in the laboratory; the latter usually inspect the whole organs and biopsy samples to decide which parts warrant detailed scrutiny and also make clinical judgements based on their

observations of the slides which have been prepared. However, suitably qualified and experienced biomedical scientists can fulfil these roles where needed.
Box 6.3 gives a list of common procedures performed by biomedical scientists
in histopathology and cytopathology laboratories.

Box 6.3   Examples of some common procedures performed in histopathology and cytopathology laboratories

Processing of tissue which has been fixed, by dehydration and embedding
   with paraffin wax

Preparing thin sections from wax embedded tissue by microtomy

Preparation of slides from fluids and swabs

Staining slides with haematoxylin and eosin (H & E) and subsequently
   more specific stains if required

Immunocytochemistry to detect specific antigens within or on the surface
   of cells

6.4.1   Diagnosis – preparing a skin biopsy for investigation of cancer

When a patient has a piece of skin removed (for example, containing a mole)
where malignancy (i.e. cancer) is suspected, it is put into a specimen container
filled with a fixative (usually formalin) straight away. This prevents the degeneration of the tissue (through autolysis and putrefaction), which would otherwise
occur as soon as it was removed from the body. The piece of skin has to remain
in formalin for at least 24 hours to ensure that it is adequately fixed. This is
necessary so that the morphology of the cells does not change during the processing stages; there is also the added advantage of 'fixing' any infectious agents
present, which makes it safe to handle tissue on the open bench (see Chapter 7).

The biopsy will then be cut into several pieces of a few millimetres thickness
and placed in plastic holder called a cassette. Each piece of tissue needs to be
filled ('impregnated') with paraffin wax to give it support during the preparation
of the thin slices which are placed on the slides. To achieve this, first the water in
the cells is removed by putting it through a series of pots containing incremen-
tally increasing concentrations of alcohol, which has the effect of dehydrating
the tissue. This is then removed ('cleared'), using a solvent such as xylene and
finally the solvent is replaced by molten paraffin wax.

The cassette is left on a cold plate to allow the paraffin wax to set, before
being taken to a microtome (which operates in a similar way to a meat slicer,

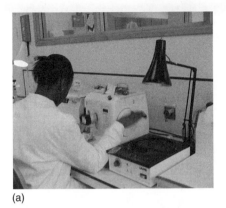
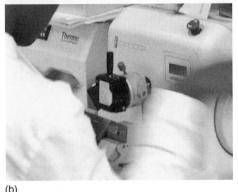

(a)   (b)

Figure 6.9 (a) A biomedical scientist cuts thin sections from tissue using a microtome; (b) close-up of the cassette containing tissue embedded in wax within a microtome

but produces much thinner slices!), where extremely thin ($3–5\,\mu$m) slices are made (see Figure 6.9). These are floated on a water bath, which makes picking them up with forceps and placing them on a glass slide easier. Floating a section on water also has the effect of flattening it, so that it sits on the slide in one plane.

The procedure of embedding with wax is then reversed, by treating the slide with xylene to remove the wax, then alcohol to remove the xylene and finally replacing the alcohol with water, so the original composition and morphology of the cells should have been restored. The slide is then stained with the standard haematoxylin and eosin ('H & E') stain, in which nuclei are stained blue and cytoplasm appears pink. Due to the large number of pieces of tissue which are processed in this way each day in a typical histopathology laboratory, machines are generally used for the dehydrating and embedding process and the dewaxing and staining step. However the microtomy is a very skilled task, which has to be done manually. A biomedical scientist will check the appearance of the stained slide, to make sure it has been prepared satisfactorily, before passing it to a histopathologist who will make a judgement on the clinical significance of the morphology of the cells and the overall structure of the piece of tissue. The slides in Figure 6.10 show the appearance of normal skin tissue and malignant melanoma.

6.4.2 Monitoring – assessing the potential efficacy of tamoxifen in treating breast cancer

Biomedical scientists and histologists also prepare and examine slides from sections of tissues and organs which have been removed because they contain malignant tumours. For example, when a discrete breast lump has been identified as cancerous, after X-ray and examination of a biopsy, the lump

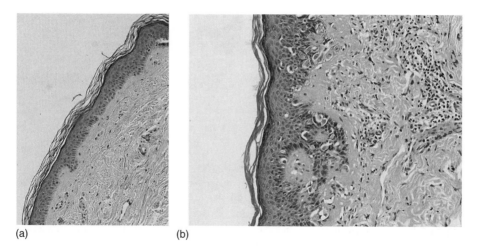

(a)   (b)

Figure 6.10 Slides prepared from skin tissue stained with haematoxylin and eosin. (a) Normal skin tissue. Note that the basal layer is well organized and the distinction between the next layer; nuclei are regularly shaped ($\times$20). (b) Skin tissue from a patient with melanoma. Note the overgrowth of the basal layer, where the basement membrane has broken; there are more nuclei, showing atypical shapes and many darker staining melanocytes, where the cells have proliferated ($\times$20). Both photographs courtesy of Ms L. Dixon

is removed surgically ('lumpectomy'). The observable cancerous area has an edge ('margin') which the surgeons take into account when excising the malignant tissue; in order to be sure that they have removed all the cells affected by the cancer, they make sure that they take out a small amount of the healthy tissue outside that margin.

The treatment for breast cancer includes radiotherapy and chemotherapy with cytotoxic drugs, intended to kill the malignant cells indiscriminately. The tumour can also be treated specifically with the drug tamoxifen, but this is not effective in all cases. It has been discovered that the drug targets particular receptors on breast cancer cells, namely the oestrogen receptor (ER) and the progesterone receptor (PR). In about one-third of breast cancer cases, the tumour cells do not have these receptors, which means that tamoxifen will not help to limit their activity. As this drug is expensive and has some unpleasant side-effects (such as hot flushes and nausea), it is better to identify which patients are likely to respond to therapy by testing for the presence of the ER and PR receptors.

This is done using the technique of immunocytochemistry, in which specific antibodies to the receptor in question are used. A slide is prepared from a piece of the cancer tissue in the same way as described for the skin biopsy above. At the stage of dewaxing prior to staining, the section is treated with a suitable solvent, such as xylene, and then heated in a microwave oven or pressure cooker, which helps to make the receptors stand out from the tissue, thus allowing specific antibodies to bind to them more easily. When ready, the slide is then incubated with the antibody raised against ER or PR, as appropriate. If there are any receptor molecules on the surface of the tumour cells, then the antibody will bind to them. This is called the 'primary antibody'. In order to visualize this reaction, a 'secondary antibody' is needed which has been designed to react with the primary antibody but also has been 'labelled' in some way to allow this reaction to be visible. A commonly used label is the enzyme horseradish peroxidase (HRP), which is also used in serological enzyme immunoassays (see Section 6.3.2). The substrate used in this context is 3,3'-diaminobenzidine (DAB), which produces a brown colour in the presence of the peroxidase enzyme. This can be clearly seen under a light microscope and thus cells which have the ER or PR receptors on their surface will appear brown when the slide is examined. The intensity and sharpness of the colour can be enhanced by using an extra step between the primary and secondary antibody; one available method exploits the high affinity that the molecules biotin and streptavidin (or avidin) have for each other. In this case, secondary antibody to which biotin has been attached, instead of the enzyme label (for example HRP), is used; thus, after the primary and secondary antibodies have reacted, there will still be no visible colour change. The enzyme used has streptavidin added; the streptavidin molecules will bind with the biotin molecules, which therefore attaches the enzyme to the antigen – antibody complex. The substrate can now be added and the colour change detected, as described above. The result from a slide prepared from breast tissue and stained with antibody against the ER receptor is shown in Figure 6.11.

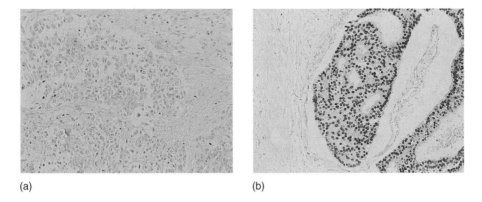

(a)   (b)

Figure 6.11 Slides showing breast tissue cells stained by immunohistochemistry for presence of oestrogen receptor. (a) Normal breast tissue. Note that the tissue contains a small number of cells staining positive for the oestrogen receptor on their surface, showing as darker stained cells (×20). (b) Breast tissue from a patient with cancer. Note the abnormally high number of cells staining positive for the oestrogen receptor, showing as darker stained cells (×20). Both photographs courtesy of Ms L. Dixon

6.4.3 Screening – cervical screening

All sexually active women over the age of 25 years in the United Kingdom are encouraged to have regular cervical swabs taken for screening, as part of a national programme which began in 1988. The aim is to identify women who are at risk of developing cancer, by noting changes in the morphology of their cervical epithelial cells, and offering them appropriate treatment in time to prevent life-threatening disease.

The method of preparing a cervical smear for examination is to make a slide, fix it in alcohol and then stain with Papanicolaou stain (the so-called 'Pap' smear). This would then be examined for abnormalities in the cells of the squamous epithelium. Each slide is then checked for the presence of changes which indicate the possibility of the patient developing cervical cancer ('precancerous' changes). The morphology of any abnormal cells and the proportion of cells on the slides which are exhibiting abnormalities signify the severity of the malignancy. The appearance is classified according to the extent of the cervical intraepithelial neoplasia (CIN) and is graded as CIN 1, CIN 2 or CIN 3, with 3 being the most serious (see Figure 6.12); the next stage is invasive cancer.

In order to process the large number of routine screening samples, laboratories employ staff in a specialized grade of associate practitioner called a 'cytoscreener'. They are rigorously trained and have to be attend refresher courses and undergo competency testing regularly. The appearance of the cells in most slides will be normal, so the cytoscreeners' role is to identify possible

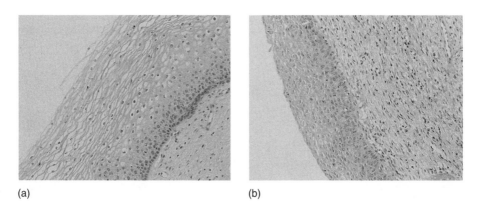

(a)   (b)

Figure 6.12 Slides prepared from cervical squamous epithelium stained with Papanicolaou stain. (a) Normal cervical squamous epithelial tissue. Note cells maturing and dying towards the outer surface of the tissue, which is a sign of normal cell programming; the distinction between the next layer is clear (×20). (b) Cervical squamous epithelium showing CIN 2. Note cells are densely packed, with more nuclei than in normal cervical epithelium and showing an irregular appearance; cells extending into the next layer of tissue indicate a loss of the normal control over the cells' activity (×20). Both photographs courtesy of Ms L. Dixon

abnormal morphology and pass those slides on to a biomedical scientist for further investigation.

The method of collecting cervical smears can markedly affect the test results. The sample may not take a representative cross-section of cells (and so miss abnormal cells); also, if there is excessive mucous in the sample or the patient has an infection, which causes inflammation and thus alters the appearance of the cells, the slides can be difficult to interpret. A change to the collection method has recently been introduced, called liquid-based cytology (LBC). This uses a plastic brush to take the cells from the cervical epithelium more evenly and this brush is placed in a pot containing a preservative fluid, which minimizes the degradation of the cervical cells in transit to the laboratory. The patient's sample is then processed in an analyser, which takes an aliquot and produces a glass slide containing an even distribution of cells. The slide is stained with the Papanicolaou stain and examined as above.

6.5 Medical microbiology

This involves the examination of a wide range of samples to look for microorganisms which cause disease (pathogens). The microorganisms of interest are bacteria, fungi, viruses, protozoa and helminths. These organisms can be identified directly by attempting to culture them, observing them microscopically or detecting the presence of a diagnostic antigen. Specimens used in microbiological culture tests include urine, faeces, blood, pus, sputum, wound swabs, genital swabs, cerebrospinal fluid, vesicle fluid, peritoneal fluid and nail clippings. Serological tests are also used to determine the presence of a microorganism indirectly, via the specific antibodies produced by the body during a particular infection and which can be found in serum. Box 6.4 gives some examples of tests commonly undertaken in a microbiology laboratory.

Box 6.4   Examples of some common tests performed in medical microbiology laboratories

Culture of bacteria and fungi from samples on agar plates

Examination of wet preparations from swabs, urine or concentrated faecal samples

Preparation, fixing, staining and examination of slides from sputum specimens, swabs, faecal samples, cerebrospinal fluid and organisms grown in culture, using Gram stain or other specific stain as appropriate

---

Box 6.4   (Continued)

Use of biochemical tests to identify organisms

Determining antibiotic sensitivities for bacteria and fungi grown in culture from clinical specimens

Serological tests for antigens and antibodies to diagnose cause of infection

Serological tests to detect antibodies which provide protection against particular infections (either through previous exposure or vaccination)

6.5.1   Diagnosis – urinary tract infection

Establishing whether a patient has a bacterial urinary tract infection (UTI) and then identification of the causative agent is important, as it allows the patient to be prescribed an effective antibiotic treatment where necessary. Due to the large volume of urine samples sent to the microbiology laboratory for diagnosis of a UTI, it is usual to conduct a preliminary screening test to ascertain that the patient has an infection. This done by estimating the concentration of white cells – either by counting the white blood cells by placing a known volume in a designated well of a 96-well microtitre plate and examining microscopically, or by using a cellulose strip with a chemical reagent which changes colour according to the white blood cell levels in the sample (see Figure 6.13).

If the white cell count is above a designated level (e.g. $>10^4$ cells/mL), then the sample is plated out on to suitable culture media, in order to grow the bacteria, thus allowing identification of the organism. Urine specimens are plated out on to a selective medium such as cytosine lactose electrolyte-deficient (CLED) agar and incubated at 37 °C overnight. Common causes of UTI include *Escherichia coli*, *Pseudomonas aeruginosa* and *Staphylococcus aureus*. Any bacteria which grow (called 'isolates') can be identified first by their effect on the CLED agar – *E. coli* produces yellow colonies which have a deeper colour in the centre, colonies of *P. aeruginosa* appear green and *S. aureus* grows as small colonies which are a uniform shade of yellow. The biomedical scientist will carry out further tests to identify the isolate fully.

It is then important to determine the antibiotics to which the organism is sensitive, which helps to ensure that the patient receives the correct treatment. For example, if the isolate was *E. coli*, it could be tested against trimethoprim and amoxycillin, which are commonly used to treat UTIs; unfortunately, many strains of *E. coli* are resistant to both of these drugs, so other antibiotics such as cephalexin and nitrofurantoin are used. Antibiotic sensitivity testing involves

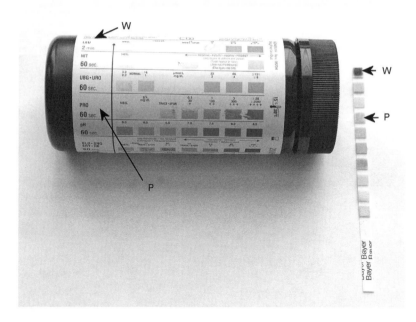

Figure 6.13 Urine dipstick test result from a patient with a urinary tract infection. Note the dark colours for the first and fourth test squares, which indicate that the urine contains abnormally high levels of white blood cells (W) and protein (P)

culturing a sample of the bacterial isolate on a plate of a special type of agar (e.g. Mueller Hinton agar) and adding small discs impregnated with the antibiotics. After overnight incubation, the biomedical scientist examines the plate; if there are areas where the bacteria have not grown close to a particular antibiotic disc, this indicates that the antibiotic has killed or inhibited the organism and would thus be suitable as a treatment (see Figure 6.14).

6.5.2 Monitoring – hepatitis B 'e antigen'/'anti-e antigen' levels

A number of markers can be found in patients who are infected with hepatitis B virus, including hepatitis B surface antigen (HBSag), antibody to hepatitis B core antigen (HB anti-core) and either hepatitis B 'e antigen' (HBeAg) or 'anti-e antigen' (anti-HBe). During the course of recovery from an acute infection, the levels of HBeAg detectable in the patient's blood should decrease, whereas the levels of the corresponding anti-HBe antibody will rise, although this often takes several months. Monitoring the course of this change from producing HBeAg to anti-HBe in a patient's serum is important as is shows that they are recovering. Some patients are unable to eliminate the virus and become chronic carriers; they may carry HBeAg in their blood or they may convert to producing anti-HBe. Blood and body fluids from a Hepatitis B positive patient

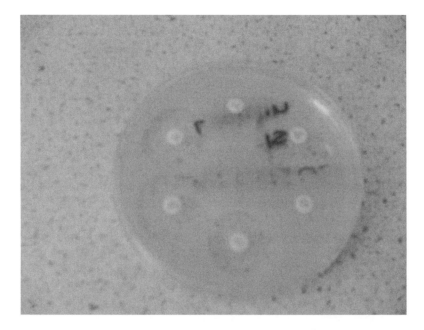

Figure 6.14 Antibiotic sensitivity plate showing an organism which is sensitive to four of the antibiotics under test (showing zones of inhibition) and resistant to the other two (no zones of inhibition)

who is producing HBeAg are more infectious and associated with a higher risk of long-term complications than those from a carrier who is producing anti-HBe.

The assay for HBeAg/anti-HBe is an enzyme immunoassay (EIA) which can detect either (or both) in the sample. This means that the biomedical scientist only has to perform one test to determine whether the patient has HBeAg or anti-HBe in their serum, although two test wells are set up for each patient's sample. The test works to detect HBeAg because the solid phase is coated with anti-HBe, so when the test serum is added and incubated under suitable conditions, any HBeAg will bind to the solid phase. This reaction is detected by adding anti-HBe conjugated to an enzyme; after incubation and washing, the substrate to the enzyme is added and this produces a quantifiable colour change. This is a simple 'sandwich' EIA. The presence of anti-HBe is detected in another well on the same plate which has again has anti-HBe as the solid phase. Before being added to the test well, the serum is mixed with a small amount of neutralizing agent which contains HBeAg. This time the reaction is a competitive EIA, whereby any anti-HBe in the sample competes with the antibody in the solid phase to bind to the free HBeAg. After incubation, the well is washed and any of the HBeAg which has bound to anti-HBe in the patient's serum will be washed away and only antigen which is attached to the anti-HBe on the solid phase will remain. After washing, the reaction is detected by adding conjugated anti-HBe as previously. In the test for HBeAg, the higher the reading, the more antigen is present in the sample; for the anti-HBe test, however, a positive result is indicated by a low reading (see Figure 6.15).

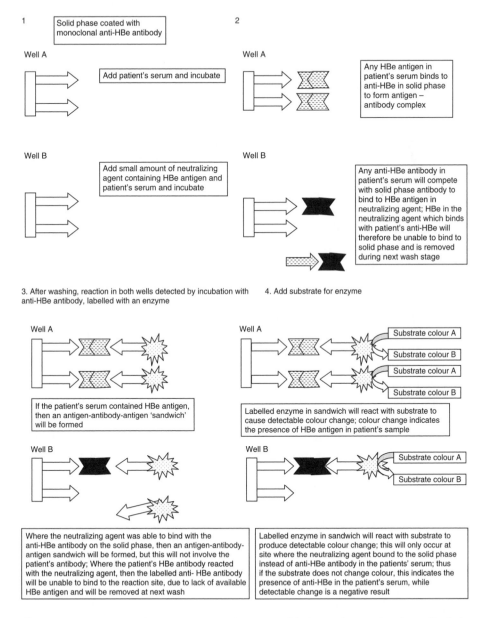

Figure 6.15 Principle of immunoassay to detect Hepatitis B e antigen and anti-e antibody in serum. In some automated analysers, whole blood is loaded on to the machine

6.5.3 Screening – national chlamydia screening programme

Chlamydia trachomatis is a sexually transmitted infection. It is estimated that between 1 in 10 and 1 in 20 of (sexually active) adults in the UK are infected with chlamydia. Infants can be also infected during birth if their mother has an active infection and be born with chlamydial conjunctivitis. Although the infection is treatable with antibiotics (tetracycline or erythromycin), the infection is often asymptomatic in adults, which means that they can pass it to partners unwittingly. The long-term consequences of untreated infection include pelvic inflammatory disease (PID) and infertility. There may be a gap of 10 or more years between a person becoming infected with chlamydia and seeking investigations to explain why they have been unable to conceive, by which time it may be too late to reverse the damage which the organism has done. Hence there is now a national screening programme for sexually active 16- to 24-year olds in the UK, which aims to detect and offer effective treatment for asymptomatic infection (www.chlamydiascreenning.nhs.uk).

Unlike tests for many other sexually transmitted infections, the screening test for *Chlamydia trachomatis* can use a mid-stream urine sample, which is easier to collect and nicer for the patient than a genital swab. People can take their own samples and home at take them to a clinic or even post them to the laboratory. Swabs taken from the genital area as part of a medical examination can also be used.

Biomedical scientists' contribution to this screening programme is through the testing of the urine or genital swabs for the presence of *Chlamydia trachomatis* antigen, using a nucleic acid amplification test (NAAT) procedure. There are three recommended NAATs, namely real-time polymerase chain reaction (PCR), strand displacement amplification (SDA) and transcription-mediated amplification (TMA). They all work on the principle of using DNA primers to target a specific area of the *Chlamydia* genome and amplify this strand. The amplicon is usually detected in similar way to an EIA, by labelling certain bases in the reaction mixture with a conjugate such as biotin, which can form the basis of an detection assay which results in a quantifiable colour change.

The test samples must be collected into specific containers. Urine must be put into a specimen pot provided as part of the test kit, which is coated with a chemical which removes anything which might interfere with the amplification step of the NAAT; similarly, genital swabs must be placed in bijou bottles filled with a specific transport medium as soon as they have been taken. The test procedure itself is fully automated, which allows large numbers of specimens to be processed quickly and also keeps the reaction area free from contamination (which can be a problem in molecular biology assays). The biomedical scientist pipettes an aliquot of each sample into a designated well in a 96-well plate and loads it on to the machine. They then read and analyse the results at the end of the run.

At the time of writing, only designated laboratories are performing the assay for the screening programme. The national guidelines about which test to use recommend that each individual laboratory consider their resources and expertise and also the suspected prevalence in their target population, as this affects the suitability of the test (see Section 6.7).

6.6 Liaison between pathology disciplines

Patients whose specimens require pathology investigations often present with common and fairly non-specific symptoms, such as pyrexia (fever), lethargy (tiredness) and purpura (rash). A range of tests might be necessary to narrow down the possible causes ('differential diagnosis') and this might involve some or all pathology disciplines. Sometimes the tests are performed simultaneously, whereas in other situations the results from one laboratory suggest that further tests are required which involve another discipline. For example, in the case of a child with suspected meningitis, cerebrospinal fluid (CSF) would be taken and aliquots sent to the biochemistry and the microbiology laboratories at the same time. The biomedical scientist in biochemistry would test for glucose and protein whereas the microbiology biomedical scientist would count the white blood cells and do specific rapid tests to identify any bacteria present (e.g. prepare a slide and stain with Gram stain). In this way, a diagnosis can be made quickly and, if the child does have bacterial meningitis, they can be given the appropriate treatment as soon as possible. However, in the case of a person infected with hepatitis, the tests might be done in stages. The symptoms that a patient with hepatitis has can sometimes be very non-specific and when they first visit their doctor, it may not be obvious. Liver function tests (LFTs) might be included in a general set of tests to find out why the patient is unwell. If the results indicated that the patient could have infectious hepatitis, then a further blood sample would be taken and sent to the microbiology (or virology) laboratory, where the exact cause can be identified.

6.7 Evaluation of a new diagnostic test

New assays are regularly produced for clinical pathology laboratories thanks to both scientific and technological breakthroughs. For example, cell surface and serological markers, which provide a more precise diagnosis or indicate a prognosis in certain cancers, are discovered intermittently. Occasionally new diseases arise which are caused by previously unknown agents, such as human immunodeficiency virus, the cause of HIV/AIDS, both of which were described for the first time in the early 1980s. Sometimes the novel finding is the source of a well-known disease; a good example of that is the recognition that the bacterium *Helicobacter pylori* is a common cause of stomach ulcers, which had previously been attributed to stress and poor diet.

In addition to scientific and clinical research, scientists are constantly working to improve test methodologies, which might provide clearer, more focused or faster results. A recent example is the liquid-based cytology (LBC) method of collection and preparation of cervical smears, discussed above in Section 6.4.3. Many diagnostic virology laboratories are replacing the traditional method of detecting viruses in samples, by culturing them in animal or human cells, with molecular methods. Before introducing a new testing method into routine use, biomedical scientists have to evaluate it carefully, to make sure that the results will be reliable and credible. This is done by comparing it against either the method which is widely accepted to be the best (the 'gold standard') or, if that is not possible, then the test which is currently in use in their laboratory.

When starting to evaluate a new test method, it is useful to consider what is required from a diagnostic test. Box 6.5 lists the main elements to examine and to compare with the existing method. The two tests should be run in parallel for a predetermined minimum number of samples (e.g. 100) or length of time (e.g. 1 month) and data collected on the performance of both, which can then be carefully compared.

Box 6.5   Key requirements of a diagnostic pathology test

A test carried out in the diagnostic pathology laboratory should:

- Be an assay for an analyte known to be of clinical value.

- Give reliable results – i.e. they are scientifically and clinically credible.

- Give reproducible results – i.e. testing aliquots of the same sample repeatedly will give the same result each time.

- Be cost effective overall – e.g. an expensive piece of equipment might save several hours in turnaround time.

- Use reagents or kits which are readily available – from a supplier who can deliver new batches rapidly.

- Produce a result within a turnaround time appropriate for the clinical condition.

- Be straightforward for the operator to perform after suitable training.

- Have a high sensitivity.

- Have a high specificity.

6.8 Sensitivity and specificity of an assay

Sensitivity is a measure of how much of the antigen (or enzyme or antibody or abnormal cell type) needs to be present in the sample before the test can detect it. Thus, the more sensitive the test, the smaller are the amounts of the antigen (or enzyme or antibody or abnormal cell type) that it can detect. However, a consequence of high sensitivity is the possibility of false positive results.

Specificity is a measure of how good the test is at finding the exact antigen (or enzyme or antibody or abnormal cell type). A very specific test will only detect the antigen (or enzyme or antibody or abnormal cell type) and nothing else. A consequence of high specificity is the possibility of false negative results.

The sensitivity and specificity of a given assay can be calculated and expressed as a percentage. The percentage sensitivity of a test is a measure of its ability to detect positives; the percentage specificity of a test is a measure of its ability to identify negatives.

When a test is applied to a large number of samples, a proportion of the results will be positive and a proportion will be negative. If the test is reasonably reliable, most of the results will actually reflect the true status of that sample. However, since no assay is 100% sensitive and specific, there will be some false positive and false negative results. Table 6.2 sets out the possible permutations.

Table 6.2 Possible outcome of test results versus true status of sample

Test result	True status	
	Positive	Negative
Positive	True positive	False positive
Negative	False negative	True negative

Sensitivity is the proportion of true positive results out of the total number of test samples which are actually positive (i.e. the number of true positives correctly detected out of all true positives actually tested):

$$\text{sensitivity} = \frac{\text{true positives}}{\text{true positives} + \text{false negatives}} \times 100$$

Specificity is the proportion of true negative results out of the total number of test samples which are actually negative (i.e. the number of true negatives correctly detected out of all true negatives actually tested):

$$\text{specificity} = \frac{\text{true negatives}}{\text{true negatives} + \text{false positives}} \times 100$$

Although some tests boast sensitivity and specificity rates of 99% or higher, no test is 100% sensitive and 100% specific. A sensitivity of 99% means that out of every 100 positive samples tested, one will be a false negative; a sensitivity of 99.9% means a false negative rate of one out of every 1000 positive samples. When evaluating a new test, it is necessary to decide on the balance between specificity and sensitivity which is required. This depends on whether a false positive or a false negative result would be more serious and more acceptable for the patient. For example, a person is more likely to accept being given a false positive cervical smear test (which was negative on the follow-up sample) knowing that true positives were not being missed. However, people might be reluctant to take an HIV test with a high false positive rate!

There are a number of factors which affect the performance of a particular test, one very important one being the 'prevalence' of the disease/condition/marker under investigation (usually expressed as the number of people per 100 000 with that disease). This affects the 'predictive value' of the test.

The positive predictive value (PPV) for a test is an expression of the probability that when the test gives a positive result, the sample is really positive, i.e. it is the proportion of positive results that are 'true' positives:

$$\text{PPV} = \frac{\text{true positives}}{\text{true positives} + \text{false positives}} \times 100$$

The negative predictive value (NPV) for a test is an expression of the probability that when the test gives a negative result, the sample is really negative, i.e. it is the proportion of negative results that are 'true' negatives:

$$\text{NPV} = \frac{\text{true negatives}}{\text{true negatives} + \text{false negatives}} \times 100$$

(referring to categories in Table 6.2).

It is particularly important to note that in situations where the prevalence is low in the population where the test will be used, the PPV can be significantly reduced.

These points are best illustrated by calculating sensitivity, specificity, positive and negative predictive values for an assay used in populations with different prevalences of the analyte in question.

Population 1: the analyte is found in 25% of the population. Test results from 1000 people are shown in Table 6.3.

Table 6.3 Results from 1000 people for an analyte which is found in approximately 25% of the population

True positive	False positive	False negative	True negative
243	22	34	701

The calculations are as follows:

$$\text{sensitivity} = \frac{\text{true positives}}{\text{true positives} + \text{false negatives}} \times 100 = 88\%$$

$$\text{specificity} = \frac{\text{true negatives}}{\text{true negatives} + \text{false positives}} \times 100 = 97\%$$

$$\text{PPV} = \frac{\text{true positives}}{\text{true positives} + \text{false positives}} \times 100 = 92\%$$

$$\text{NPV} = \frac{\text{true negatives}}{\text{true negatives} + \text{false negatives}} \times 100 = 95\%$$

Population 2: the analyte is found in 1% of the population. Test results from 1000 people are shown in Table 6.4.

Table 6.4 Results from 1000 people for an analyte which is found in approximately 5% of the population

True positive	False positive	False negative	True negative
33	27	21	919

The calculations are as follows:

$$\text{sensitivity} = \frac{\text{true positives}}{\text{true positives} + \text{false negatives}} \times 100 = 61\%$$

$$\text{specificity} = \frac{\text{true negatives}}{\text{true negatives} + \text{false positives}} \times 100 = 97\%$$

$$\text{PPV} = \frac{\text{true positives}}{\text{true positives} + \text{false positives}} \times 100 = 55\%$$

$$\text{NPV} = \frac{\text{true negatives}}{\text{true negatives} + \text{false negatives}} \times 100 = 98\%$$

Quick quiz

1. List the main pathology disciplines.

2. Give three examples of tests commonly performed in haematology laboratories.

3. Give three examples of tests commonly performed in clinical biochemistry laboratories.

4. Give three examples of procedures commonly carried out in histopathology and cytopathology laboratories.

5. Give three examples of tests commonly performed in microbiology laboratories.

6. Name two diagnostic tests which are based on the sandwich EIA principle.

7. Explain how either aperture impedance or light scatter are used in the counting of cells in a blood sample.

8. Why must the clotting times for patients on anticoagulant therapy be regularly monitored?

9. Can a person with blood type A Rh+ donate blood to a person whose blood type is O Rh−? Explain the reasons for your answer.

10. Explain the principle of agarose gel electrophoresis.

11. The blood concentration of which hormone is measured to monitor the effectiveness of thyroxine treatment?

12. What are the possible long-term complications of uncontrolled diabetes?

13. Why are tissue samples embedded with paraffin wax before thin sections are cut from them?

14. How is this process of embedding reversed within the thin sections once they have been placed on a slide?

15. What do the abbreviations 'H' and 'E' stand for in the name of a common histology stain and why are these chemicals needed in the stain?

16. What does the abbreviation 'CIN' stand for?

17. What is CLED agar?

18. Is a person who carries hepatitis B e antigen in their blood more or less infectious than a person with anti-HBe?

19. Which age group is specifically targeted in the United Kingdom chlamydia screening programme?

20. How would you calculate the sensitivity of a test assay?

21. If a test was 99.9% specific, how many test results out of each 1000 would be false negatives?

22. Define the negative predictive value (NPV) of a test in a given population.

23. How is the positive predictive value (PPV) of a test assay for a certain disease affected by the prevalence of that disease in the population?

Suggested exercises

1. Compile a list of all the tests performed in the laboratory that you are currently working or training in and find out the reference ranges or normal results for each test.

2. Find an example of a test which is commonly performed in the laboratory that you are currently in that is usually done at the same time as tests are being carried out on samples from the same patient in other departments; write an essay to explain how this benefits the patient.

3. Liaise with your Training Officer or other senior colleague to choose a condition where more than one pathology discipline would be involved in the diagnosis of the illness and/or monitoring of the patient. Investigate which samples would be collected from the patient, which tests would be performed in each laboratory and what the expected results would be. Collate your findings in the form of a poster.

Suggested references

Crocker, J., and Burnett, D. (eds) (2005). *The Science of Laboratory Diagnosis*, 2nd edition. Chichester: John Wiley and Sons, Ltd.
Biomedical sciences explained series:
Cook, D. J. (2006). *Biomedical Sciences Explained: Cellular Pathology*, 2nd edn. Bloxham, Oxfordshire: Scion Publishing.
Hannigan, B. M. (2008). *Biomedical Sciences Explained: Immunology*, 2nd edn. Bloxham, Oxfordshire: Scion Publishing.
Luxton, R. (2008). *Biomedical Sciences Explained: Clinical Biochemistry*, 2nd edition: Bloxham, Oxfordshire: Scion Publishing.
Overfield, J., Dawson, M., and Hamer, D. (2007). *Biomedical Sciences Explained: Transfusion Science*, 2nd edn. Bloxham, Oxfordshire: Scion Publishing.
Pallister, C. J. (2008). *Biomedical Sciences Explained: Haematology*, 2nd edn. Bloxham, Oxfordshire: Scion Publishing.

Gosling, P. J. (2002). *Dictionary of Biomedical Sciences*. London: Taylor and Francis.

Haaheim, L. R., Pattison, J. R., and Whitely, R. J. (eds) (2002). *A Practical Guide to Clinical Virology*, 2nd edn. Chichester: John Wiley and Sons, Ltd.

Struthers, J. K., and Westran, R. P. (2005). *Clinical Bacteriology*. London: Manson Publishing.

Wood, J. (2004). Technical validation of a diagnostic test. *Biomedical Scientist*, **48**, 34–36.

Wood, J. (2004). Use of a diagnostic test in clinical practice. *Biomedical Scientist*, **48**, 149–151.

7
Biomedical science laboratory techniques

7.1 Introduction

Biomedical scientists must demonstrate competency in a wide range of laboratory techniques. In many instances, this can be achieved by describing the procedures applied during the passage of a specimen through the diagnostic laboratory (including sample handling, health and safety issues, instrumentation, methodology, reagent preparation, quality control, result interpretation and possible errors). Within each pathology discipline there are some specialist techniques which are largely or exclusively used in a particular laboratory. In addition, there are a number of generic laboratory skills and techniques practised across specialist disciplines and by graduate laboratory bioscientists in general for which evidence of competency is required towards the IBMS Registration Portfolio (Table 7.1). The aim of this chapter is to describe some of the basic techniques with which all biomedical scientists are expected to be familiar, together with a selection of advanced techniques used in the main specialist discipline laboratories. It should be possible to carry out the techniques described in the sections below in most standard pathology specialist discipline laboratory settings. In some laboratories, the automation of many routine clinical measurements, particularly in haematology and clinical chemistry, may restrict the opportunities for students to gain experience in certain techniques. Accordingly, examples of 'hands-on' exercises are included which allow students to carry out techniques and associated applications that would normally be part of the automated process. In addition to providing students with the laboratory skill itself, a 'manual' method can be of considerable value in enhancing the understanding of the principles and limitations of methodologies employed. Exercises in basic skills and data interpretation are linked to the main exercises, which may be

An Introduction to Biomedical Science in Professional and Clinical Practice   Sarah J. Pitt and James M. Cunningham
© 2009 John Wiley & Sons, Ltd

Table 7.1 Generic and discipline-specific techniques required by biomedical scientists[a]

Generic

Specimen handling[b]
Liquid handling[b]
Solution preparation[b]
Microscopy[b]
Spectroscopy[b]
Centrifugation[b]
Electrophoresis[b]
Chromatography[b]

Electroanalytical techniques
Immunological techniques
Enzyme assay[b]
Molecular biology techniques
Work with accuracy and precision[b]
Calibration and quality control checks[b]

Haematology and transfusion science

Blood and bone marrow smear preparation and staining
Cell morphology
Full blood count[b]
Blood grouping
Coagulation
Compatability testing
Serological methods
Blood components processing and monitoring

Microbiology and virology

Sterilization and disinfection
Handling culture medium[b]
Sample inoculation and sub-culturing[b]
Culture methods for identification of common pathogens
Susceptibility testing
Serological procedures for common pathogens
Microscopic identification of bacteria, parasites, fungi or viruses

Histopathology and cytology

Tissue and cell preparation[b]
Microtomy[b]
Cytological staining
Histological staining for specified cell and tissue components[b]

Clinical biochemistry

Liver function tests[b]
Assay of:
    Cardiac markers[b]
    Urea and electrolytes
    Hormones
    Proteins
    Carbohydrates[b]
Therapeutic drug monitoring

Immunology

Serological antigen/antibody reactions;
Immunological assay
Complement assay
Immunophenotyping

---

[a]The table lists laboratory procedures for which competency is required by the IBMS Certificate of Competence Registration Training Portfolio; these can be classified broadly into 'generic' techniques, with which all biomedical scientists might be expected to be familiar, and those specific to investigations carried out in specialist discipline laboratories.
[b]Aspects of the techniques indicated are covered by exercises described in this chapter.

helpful to students first encountering the basic laboratory skills in the placement laboratory and, for others, to reinforce and consolidate any previous experience. In addition to the guidance provided to carry out the selected techniques successfully, it is also important that local standard operating and health and safety procedures are followed (see Chapter 5), and students should always consult with laboratory supervisors if unsure about any aspect of this.

7.2 Haematology

7.2.1 Haemoglobin measurement

Measurement of blood haemoglobin is one of the parameters of the full blood count, which is one of the most routinely requested tests in haematology. Its measurement is used towards the detection of diseases that cause a deficiency or excess of haemoglobin, such as anaemia or polycythaemia. Although several methods exist for its estimation, the cyanmethaemoglobin method recommended by the International Committee for Standardization in Haematology (ICSH) is widely used manually and in automated analysers. The cyanmethaemoglobin method is based on the oxidation of iron in haemoglobin from lysed red cells by potassium ferricyanide to form methaemoglobin; reaction of methaemoglobin with potassium cyanide forms cyanmetahaemoglobin, which absorbs light at 540 nm.

7.2.1.1 Materials

Blood sample (EDTA anticoagulated)
Cyanmethaemoglobin standard solution (ICSH preparation BS3985)
Drabkin's reagent (Van Kampen and Zijlstra modification):

Potassium ferricyanide	0.200 g
Potassium cyanide	0.050 g
KH_2PO_4	0.140 g
Made up to 1 L in distilled water	
pH 7.0–7.4	

Visible range spectrophotometer

7.2.1.2 Method

1. Pipette the solutions listed in Table 7.2 into 15 mL test-tubes (in duplicate). Ensure the blood sample is well mixed before adding (e.g. using a mechanical agitator for 3–5 minutes or gentle inversion in tube 20–25 times).

2. Mix the contents of the tubes thoroughly and leave to stand for 10 minutes.

Table 7.2 Solutions for haemoglobin measurement

Solution	Cyanmethaemoglobin standard (A–F) or sample (G) (μL)	Drabkin's reagent (μL)	Hb (g/dL)[a]
A	–	5000	0
B	1000	4000	3.01
C	2000	3000	6.02
D	3000	2000	9.04
E	4000	1000	12.05
F	5000	–	15.06
G	20	5000	From calibration curve

[a] Assuming ICSH standard = 0.06 g/dL; concentration shown corrected for dilution factor applied to measurement of sample (20 μL added to 5000 μL = ×251 dilution factor).

3. Place a cuvette containing Drabkin's solution alone (tube A) in the spectrophotometer and set the reading to zero at a wavelength of 540 nm.

4. Measure the absorbance of the standard solutions B–E and the dilution of the 'unknown' sample G.

7.2.1.3 Results and analysis

Using absorbance data from standard solutions A–E, plot a calibration curve of absorbance on the y (vertical) axis versus concentration of haemoglobin (g/dL) on the x (horizontal axis); use the absorbance values obtained from solution F to estimate the concentration of haemoglobin in blood sample Y.

Normal range values:

Men	13.5–18.0 g/dL
Women	11.5–16.5 g/dL
Infants	13.5–19.5 g/dL
Children	10.0–14.8 g/dL

By carrying out this discipline-specific measurement, this exercise provides experience in the generic techniques of making up and diluting solutions, pipetting, spectrophotometry and determination of a sample concentration by calibration curve analysis. Using pipettes to dispense liquids with precision and accuracy is one of the essential basic laboratory skills on which many procedures in the bioscience laboratory have a fundamental reliance for accurate and precise volume measurements. Accordingly, it is very important that a laboratory biomedical scientist is competent in the basic skill of measuring liquid volumes with accuracy and precision. An associated exercise in basic pipetting skills including estimation of precision and accuracy is described in Box 7.1.

Box 7.1 Precision and accuracy of pipetting

The precision and accuracy of pipetted volumes can be conveniently estimated by repeated measurement of the mass delivered of a liquid of known specific gravity.

Precision is a measure of the 'spread' of the data obtained from repeated measurements and can be expressed as a **coefficient of variation value (% CV)**, which is calculated using the **mean** and **standard deviation (SD)** values for each set of measurements as follows:

$$\% \, CV = (SD/mean) \times 100$$

The precision of delivery from an automatic pipettor of capacity 200–1000 μL can be estimated by carrying out the following exercise:

1. Weigh a suitable container (e.g. a plastic weighing boat) using an analytical balance accurate to 0.001 g.

2. Record the container mass or set the balance to zero using the 'tare' function.

3. Pipette 1000 μL aliquots of distilled water into the container and record the mass of the volume dispensed. Repeat the measurement, e.g. for 10 aliquots dispensed. Repeat the procedure using aliquots of 500 and 200 μL.

4. Calculate the % CV from the values for the mean and SD of each set of measurements at a single volume.

The level of precision for a particular measurement which is acceptable depends on the requirement of the particular procedure, but for routine measurements in a bioscience or diagnostic laboratory a value of 0.5–0.1% can be an acceptable threshold.

Accuracy is defined by the International Standards Organization (ISO) as 'the closeness of the agreement between a test result and the accepted reference value'. There is not always an accepted reference value to compare against, but in the case of the pipetting exercise described above, if the density of water is taken to be 1.000 g/mL then the accuracy is quantified by $x - u$, where x = mean weight measured of pipetted volume and u = expected weight of water of volume 1000, 500 and 200 μL. A systematic error in the procedure giving measured values that are consistently less than expected (a negative measurement bias) or greater than expected (a positive measurement bias) is indicated by a positive and negative value for $x - u$, respectively.

Box 7.1 (Continued)

To test whether any bias is statistically significant, a calculation of the *t*-statistic can be carried out (assuming a normally distributed population of measurement values) as follows:

$$t = (x - u)\sqrt{n}/\text{SD}$$

where x = sample mean, u = expected or known standard value, SD = sample standard deviation and n = sample size.

The *t*-value is compared against a chosen critical value for significance as shown in standard statistical tables. If *t* exceeds the chosen critical value then the null hypothesis 'there is no difference between the observed value (x) and the expected value (u)' is rejected, i.e. there is judged to be a difference between the two values.

7.2.2 Red cell count

Another key parameter of the full blood count is the estimation of the number of red cells in a unit volume. Although this is routinely carried out using electronic counting apparatus, which has much greater convenience and reduced associated errors, a manual count may be carried out as a reference method. For the manual method, a specialized counting chamber is used, which is an item of precision glassware that allows the number of cells contained within a specified volume to be accurately counted. Different types of chambers exist, but one commonly used and which is recommended for blood cell counts is the Improved Neubauer chamber. Manual counting of white blood cells and platelets can also be carried out in a similar way using a counting chamber with use of appropriate solutions and dilutions.

7.2.2.1 Materials

Blood sample
Formol – citrate solution (32 g/L trisodium citrate containing 10 mL/L of 40% formaldehyde)
Improved Neubauer counting chamber (BS 748)
Light microscope ($\times$40 magnification)

7.2.2.2 *Method*

Mix blood well and dilute 20 μL in 4 mL of formol – citrate solution.
Prepare the counting chamber and fill it with diluted blood sample as follows:

1. Wash the chamber and cover slip using distilled water followed by alcohol and dry.

2. Place chamber on a firm, flat surface and press the cover slip over the central counting area; the appearance of coloured rings of light (Newton's rings) indicates that the cover slip is flush with the surface of the slide and a chamber of the correct volume is present for the cell count,

3. Fill the chamber by allowing the mixed diluted blood to flow under the edge of the cover slip, for example using a glass capillary tube or micropipette. Clean the slide and cover slip and repeat the procedure in the event of any of overflow of fluid into the moat, incomplete filling of the chamber or formation of air bubbles in the chamber.

4. Allow the cells to settle for 2–3 minutes free from direct sun, a heat source or vibrations.

5. Using a ×40 objective, count cells in five of the 25 smaller squares in the central ruled area of the chamber (0.02 μL).

7.2.2.3 *Calculation*

The number of cells in a unit volume can be calculated using the following equation:

cell count (cells/L) = N × 1/unit volume (L) ×dilution

For the example method above (×201 dilution, 0.02 μL unit volume) and taking an example count of 520 cells:

$$\text{red cell count} = 520 \times \frac{1}{2 \times 10^{-8}\,\text{L}} \times 201 = 5.2 \times 10^{12}/\text{L}$$

Normal range values:

Men	$4.5–6.5 \times 10^{12}/\text{L}$
Women	$3.8–5.8 \times 10^{12}/\text{L}$
Infants (newborn)	$4.0–6.0 \times 10^{12}/\text{L}$
Children (3 months)	$3.2–4.8 \times 10^{12}/\text{L}$
Children (1–12 years)	$3.6–5.4 \times 10^{12}/\text{L}$

The manual red cell count exercise requires solution preparation, accurate dilutions of blood to be made and careful use of a precision cell counting chamber viewed under a light microscope. Manual cell counts are an acceptable alternative to automated methods provided that the number of cells counted is sufficient for the inherent error to be acceptably low; an exercise to assess the statistical error value of count data is described in Box 7.2.

Box 7.2   Statistics of count data

In all cell counting measurements, there is an inherent random error due to the random distribution (Poisson distribution) of particles. Calculation of the error for a given count value N allows a level of statistical confidence to be attached to the measurement. This is derived using the **standard deviation of the mean (SD)** of the **count N**, as follows:

$$SD = \sqrt{N}$$

A **confidence interval for the count (CI)** is calculated from the *t*-statistic given for the chosen level of significance:

$$CI = SD \times t$$

For example, to calculate a **95% confidence interval** using the example given above for the red cell count exercise where 520 cells were counted:

$$CI = \sqrt{520} \times 1.96 = 44.7$$

i.e. there is a 95% chance that the true value of the concentration of cells in the sample applied to the counting chamber lies between 520 ± 44.7 cells per unit volume ($0.2\,\mu L$) used for the count (or, expressed as a range, 475–565 cells per unit volume). Following the calculation for the red cell count, this gives a 95% confidence interval value of 4.75–5.65×10^{12} cells/L.

The **precision** of the count can be calculated by $(SD/N) \times 100$; in this example:

$$precision = 22.8/520 = 0.044$$

(or, expressed as a percentage, 4.4%).

The precision of the estimate can be increased by counting more cells; by a reverse of the calculation steps shown above for a **chosen threshold value for precision ($z\%$)**, a **minimum count number (N)** that will achieve this level of data spread can be found:

$$N = (100/z\%)^2$$

For example, the number of cells required to be counted to achieve a precision of 1% would be:

$$N = (100/1)^2 = 10000\,cells$$

7.3 Clinical chemistry

7.3.1 Serum lactate dehydrogenase activity

Lactate dehydrogenase (LDH) is a cytosolic enzyme which catalyses the interconversion between lactate and pyruvate as follows:

$$CH_3CH(OH)COO^- + NAD \rightarrow CH_3C(O)COO^- + NADH + H^+$$
$$\underset{\text{Lactate}}{} \qquad \underset{\text{Pyruvate}}{}$$

Release of LDH from damaged or lysed cells increases the enzyme activity in the circulating plasma and its measurement can be a valuable diagnostic indicator for clinical conditions, including myocardial infarction, hepatitis and certain malignancies. The exercise described here is a good example of a diagnostic enzyme assay which provides students with further experience in solution preparation and spectrophotometry. Serum preparation includes the use of a centrifugation step, which is a technique used in most standard laboratory settings; an associated centrifugation exercise to give students further experience in this technique is described in Box 7.3.

7.3.1.1 Materials

Tris – EDTA buffer (pH 7.4):

Tris(hydroxymethyl)aminomethane   6.8 g
EDTA disodium salt dihydrate   2.1 g
Make up to 1 L with distilled water

NADH solution (0.17 mmol/L):

NADH disodium salt trihydrate   13 mg
Dissolve in 100 mL of Tris – EDTA buffer

Sodium pyruvate solution (13.5 mmol/L): 149 mg

Dissolve in 100 mL of distilled water

Serum sample:

Allow whole blood to clot and centrifuge to separate serum from pellet and particulate matter. Care needs to be taken with this stage as erythrocytes contain LDH, so haemolysed serum will therefore produce erroneous values

UV – visible spectrophotometer, with thermostatically controlled cuvette holder

7.3.1.2 Method

1. Place 2.0 mL of Tris – EDTA, 0.05 mL of serum and 0.1 mL of NADH solution in a cuvette and mix.

2. Allow to stand for 5–15 minutes (37 °C); absorbance readings should be stable prior to addition of pyruvate.

3. Add 0.2 mL sodium pyruvate solution pre-warmed to 37 °C and record the absorbance at 340 nm for 5 minutes; take readings at 1 minute intervals (non-linear results may be found when the enzyme activity is high, and when this is found the serum can be diluted in phosphate-buffered saline and the assay repeated).

7.3.1.3 Results and analysis

Serum lactate dehydrogenase activity $(IU/L) = \Delta A_{340}/6.3 \times 1000 \times 2.25/0.05$ 1 IU (International Unit) of enzyme activity transforms 1 μmol of substrate per minute under defined conditions; ΔA_{340} (average change in absorbance per minute) is obtained from a plot of change in absorbance (y axis) against time (x axis).

Box 7.3   Centrifugation exercise

For cells suspended in a given medium and subjected to a particular gravitational force, the rate of sedimentation is affected by their size, density and shape. For most biological separations, the rate of separation of cells (or smaller components) is too slow under normal gravitational force and centrifugation is applied to accelerate this process. The force applied by centrifugation is usually expressed as a multiple of the acceleration due to gravity, known as the relative centrifugal force (RCF or g value). The RCF is proportional to the distance between the rotational axis and the point of rotation (r) and the speed of rotation (revolutions per minute; rpm) in the following way:

$$RCF = (1.11 \times 10^{-5})(rpm)^2 r$$

As there are many models of centrifuge available in which differ in rotational speed (usually variable within a particular range) and radius (usually fixed in a particular centrifuge by the design of the rotor used), it often preferable to state the centrifugation conditions as RCF rather than simply as rpm.

Materials and equipment required

Sample of blood (EDTA anticoagulated, e.g. five samples of 3–4 mL). Bench centrifuge (capable of $2100g$).

Automated blood cell analyser (e.g. Sysmex, Coulter); alternatively, a method can be used from a standard haematology laboratory SOP or laboratory text for the manual determination of platelet concentration using a counting chamber.

Procedure

From each sample of blood, aliquot 3×1 mL into clear plastic centrifuge tubes; keep an additional sample of whole blood for blood cell analysis (the volume required may vary according to the analyser available but typically would be less than $100\,\mu L$). Centrifuge 1 mL blood samples for 6 minutes each at $380g$, $1100g$ and $2100g$ (you may need to check the radius of the rotor in order to calculate the rpm required to produce the different RCF required). Carefully remove the plasma supernatant into a separate plastic tube and determine the platelet concentration in the whole blood sample and each supernatant. If sufficient samples are available for replicate determinations, calculate the mean and standard deviation of the platelet count for whole blood and plasma sample and assess the effect of each step change in centrifugal force applied (i.e. whole blood versus $380g$; $380g$ versus $1100g$; $1100g$ versus $2100g$. The results should demonstrate the importance of using appropriate centrifugation conditions to achieve platelet-poor plasma, which is essential for accurate clotting screen determinations (e.g. PT and APTT).

7.3.2 Analysis of urine sugars by thin-layer chromatography

Raised levels of urine glucose are a well-known indicator of diabetes mellitus. However, a number of other reducing sugars can be detected in urine and their detection can be important in diagnosis. Chromatographic separation allows the identification of closely related compounds and the method described below depends on the relative partitioning of sugars between a mobile phase and a stationary phase in the form of a thin layer (e.g. 250 μm) of silica spread on a supporting inert surface such as a glass plate, plastic or metal foil. The method can be carried out using relatively simple apparatus and materials and this exercise in chromatography provides students with further experience in solution preparation and small-volume liquid handling.

7.3.2.1 Materials

Mobile phase:

n-Butanol	75 mL
Glacial acetic acid	25 mL
Water	6 mL

Stationary phase:

If carried out routinely, a laboratory may have a method and apparatus for preparing thin-layer plates 'in-house'. However, if not available, suitable plates can be obtained commercially, e.g. silica gel 60

Solvent tank:

Glass solvent tank, e.g. 12 × 12 × 6 cm with glass lid

Standards:

Glucose, galactose, fructose, xylose, lactose dissolved in 10% mobile phase solvent

Developing reagent:

Aniline	1 mL
Diphenylamine	1 g
Acetone	100 mL
85% phosphoric acid	10 mL

7.3.2.2 Method

1. Apply 2 μL of sample using a suitably small-volume capillary tube, syringe or micropipette on to a chromatography sheet. The diameter of the spot should be kept as small as possible by applying a single drop at a

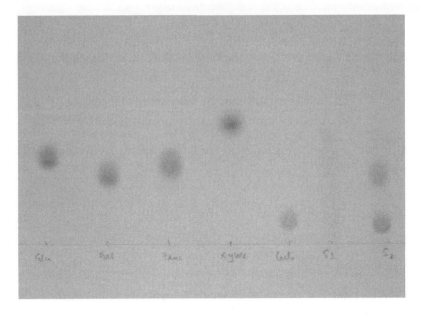

Figure 7.1 Thin-layer chromatogram showing separation of reducing sugars. From left to right: lane 1, glucose; lane 2, galactose; lane 3, fructose; lane 4, xylose; lane 5, lactose; lane 6, 'negative' urine sample; lane 7, urine sample containing added lactose and fructose. Standard concentrations 0.5% (w/v)

time and drying the spot carefully between applications, e.g. using a hair dryer. The sample spots should be applied along a line of origin which is parallel with one axis of the sheet and which is at least 2 cm from the bottom edge; there should be a minimum distance of approximately 1 cm between samples and between the outlying samples and the sheet edge perpendicular to the line of origin (Figure 7.1).

2. Pour solvent into the solvent tank to a depth of approximately 0.5 cm and place the sheet in the solvent with the samples nearest to the solvent, taking care that the solvent surface is below the line of origin.

3. Put the solvent tank lid in place and leave for approximately 2 hours or until the solvent front has moved at least three-quarters of the length of the sheet.

4. Remove the sheet from the tank and mark with a pencil the distance travelled by the solvent front. Dry the sheet with a hair dryer or by leaving overnight in a fume cupboard.

5. Apply the developing reagent evenly over the surface of the plate using an atomizing spray bottle and heat for a few minutes at 95–100 °C. Sugars are identified by blue – green spots.

7.3.2.3 Results and analysis

Measure the distance travelled by the sample spots and calculate the R_f value as follows:

$$R_f = \frac{\text{distance travelled by substance from origin}}{\text{distance travelled by solvent front from origin}}$$

The R_f values can be compared with the known values for reducing sugars as follows:

Lactose	0.14
Galactose	0.35
Fructose	0.39
Glucose	0.42
Pentoses	0.5–0.9

An example result of a separation of standard sugars using this method is shown in Figure 7.1.

7.4 Medical microbiology

7.4.1 Concentration of viable bacterial cells in broth culture

It is sometimes useful to have a reasonably accurate figure for the concentration of live ('viable') bacterial cells in a clinical sample of body fluid, liquid broth culture or suspension prepared from a colony. The following method of estimating the viable count of bacteria in an overnight broth culture is based on the Miles and Misra method:

1. Prepare an overnight broth culture of the laboratory control plate of *Escherichia coli* or *Staphylococcus aureus*, by inoculating a loopful into 20 mL of nutrient broth and incubating at 37 °C for 18–24 hours.

2. Mark up eight nutrient agar plates into three sectors, by drawing lines on the base of the plate (not the lid!) using a permanent marker.

3. Prepare serial 10-fold dilutions of the overnight broth culture as follows:

   a. Place six test-tubes in a rack labelled 1–6. Into tubes 2–6 pipette 9 mL of sterile phosphate-buffered saline (PBS).

   b. Pipette 2 mL of the overnight broth culture into test-tube 1.

    c. Take 1 mL of the culture from tube 1 and pipette it into tube 2. Mix by pipetting up and down several times and then finally take up 1 mL and transfer it into tube 3.

    d. Repeat this mixing and transfer for tubes 4–6. Take up 1 mL from tube 6 after mixing in the same way as for the other tubes and discard.

    e. The dilution of the broth culture in each tube will therefore be neat, 10^{-1}, 10^{-2}, 10^{-3}, 10^{-4}, 10^{-5}.

4. Label the agar plates: neat, 10^{-1}, 10^{-2}, 10^{-3}, 10^{-4}, 10^{-5}, broth control, PBS control.

5. For the undiluted broth culture ('neat'), pipette $100\,\mu L$ into each of the three sectors on the plate (i.e. three replicates). Repeat for the five dilutions. Use an uninoculated aliquot of nutrient broth from the same batch used in the experiment and the bottle of PBS used to make the dilutions for the control plates.

6. Allow the drops of liquid to dry and then incubate the plates at $37\,°C$ overnight.

7. The following day, examine the plates for growth (and check that there was no growth in the control plates).

8. Select the first dilution at which it is possible to count 200 or fewer discrete colonies and count the number of colonies in each of the three sectors. Calculate the mean count of colony-forming units (cfu).

9. Using the known dilution factor and the original volume of diluted broth pipetted on to the plate in each sector (i.e. $100\,\mu L$), calculate the cfu/mL for the original overnight broth culture.

10. Compare your finding with the generally used assumption that an overnight broth culture of bacteria contains 10^9 cfu/mL

7.4.2 Total number of bacterial cells in broth culture

It is sometimes necessary to estimate the total number of bacterial cells in a suspension, regardless of whether they are viable. In this case, culture methods are not suitable, but a counting chamber can be used.

1. Prepare a broth culture of *Escherichia coli* (or other Gram-negative bacillus) by inoculating a loopful from the culture plate into 20 mL of nutrient broth.

2. Incubate at $30\,°C$ for 36–48 hours.

3. Take a 20 mL universal bottle and add 10 mL of the broth culture and 1 mL of sterile PBS (a 1 in 10 dilution). Add 2–3 drops of formaldehyde (to prevent further bacterial growth) and mix thoroughly.

4. Prepare the counting chamber and fill it with diluted broth culture sample as follows:

   a. Wash the chamber and cover slip using distilled water followed by alcohol and dry.

   b. Place chamber on a firm, flat surface and press the cover slip over the central counting area; the appearance of coloured rings of light (Newton's rings) indicates that the cover slip is flush with the surface of the slide and a chamber of the correct volume is present for the cell count.

   c. Fill the chamber by allowing the mixed, diluted broth to flow under the edge of the cover slip, for example using a glass capillary tube or micropipette. Clean the slide and cover slip and repeat the procedure in the event of any of overflow of fluid into the moat, incomplete filling of the chamber or formation of air bubbles in the chamber.

   d. Allow the cells to settle for 2–3 minutes free from direct sun, a heat source or vibrations.

   e. At $\times 100$ magnification, count the cells in five of the 25 smaller squares in the central ruled area of the chamber ($0.02\,\mu L$).

5. The number of cells in a unit volume can be calculated using the following equation:

cell count (cells/L) $= N \times 1/$unit volume (L) $\times$ dilution

7.5 Histopathology

The exercises in this section are designed to illustrate some principles of fixation and staining procedures and to provide students with experience in histological techniques including tissue fixation, processing and sectioning, reagent preparation and staining for microscopic examination and evaluation of results.

7.5.1 Haematoxylin – eosin staining

Haematoxylin is a natural basic dye that stains cell nuclei. There are many methods that employ haematoxylin as a general nuclear stain, for example with eosin as a cytoplasmic counterstain to demonstrate the general state and architecture of cells and tissues. It can also be used in combination with other stains to aid the highlighting of specific cell and tissue components. Haematoxylin has, by itself, only poor staining properties and is usually used in conjunction with a mordant such as alum or iron salts which have affinity both for tissue component and the stain. The exercise described demonstrates the importance of an alum mordant to the staining procedure by comparison of results using staining solutions prepared with and without alum.

7.5.1.1 Materials

Paraffin tissue sections, e.g. liver or gut
Haematoxylin solution A (Mayer's haematoxylin):

Ammonium alum	5 g
Chloral hydrate	5 g
Haematoxylin	0.1 g
Citric acid	0.1 g
Sodium iodate	0.02 g
Distilled water	100 mL

Dissolve haematoxylin in water with aid of gentle heat followed by sodium iodate and ammonium alum; add citric acid and chloral hydrate.

Haematoxylin solution B:

As solution A but omitting ammonium alum

Scott's tap water substitute:

Sodium bicarbonate	3.5 g
Magnesium sulfate	20 g
Tap water	1000 mL

Eosin solution:

1% (w/v) eosin in distilled water

7.5.1.2 Method

1. Stain the dewaxed and hydrated tissue section in solution A or B for 5 minutes.

2. Wash well in Scott's tap water substitute until a blue tinge appears (approximately 5 minutes).

3. Stain in eosin solution for 2 minutes.

4. Wash in tap water for 5–10 seconds.

5. Dehydrate the section by taking through an alcohol series; clear in solvent.

6. Mount the slide and view under a microscope.

7.5.1.3 Results and analysis

The cell and tissue components under haematoxylin – eosin staining can be identified by the characteristic staining of nuclei (blue), red blood cells (red) and muscle/connective tissue/cytoplasm (shades of pink). To compare and report the staining results, randomly select five fields of view using $\times 10$– $\times 40$ objectives and assess the intensity of staining of selected section components from slides treated with haematoxylin solution A and B. By paying particular attention to nuclear staining and chromatin detail, the importance of the mordant in forming a link between the stain and tissue components should be readily apparent (Figure 7.2).

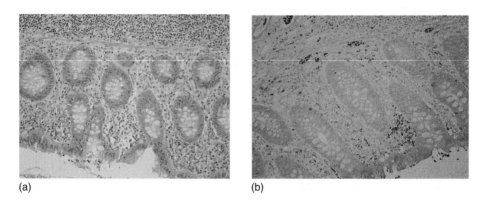

(a)   (b)

Figure 7.2 Light micrographs of haematoxylin – eosin-stained human appendix sections ($\times 20$): (a) tissue stained using standard preparation of Mayer's haematoxylin; (b) tissue stained with Mayer's haematoxylin prepared minus alum mordant

7.5.2 Martius Scarlet Blue staining

Fixation of tissue is a critical step which prevents tissue degradation and maintains architecture and composition as closely as possible to the living state. The effectiveness of a subsequent staining procedure can depend on the specific chemical and structural effects of a particular fixative and usually particular fixatives will be recommended for specific investigations in cellular pathology. This exercise is designed to demonstrate the effects of different fixatives (formol saline, alcoholic formaldehyde and Bouin's fluid) on the staining with Martius

Scarlet Blue (MSB) of a selected tissue type. MSB is an example of a trichrome stain that can be used for demonstration of the protein fibrin and can be used in the diagnosis of connective tissue disorders. The differential staining effect of MSB is believed to rely on the molecular sizes of the staining molecules and the relative tissue permeability. Formaldehyde-based solutions are the most widely used fixatives in histopathology and are commonly used with trichrome stains such as MSB. Alcoholic formaldehyde can be used as a tissue fixative for investigation of tissue glycogen and protein and Bouin's fluid is a picric acid-based fixative used to enhance staining of acidic structures, e.g. mitotic figures in dividing cells; both fixatives can, however, cause marked tissue shrinkage and distortion.

7.5.2.1 Materials

Paraffin tissue sections, e.g. placenta, kidney, liver

Formol saline:

40% formaldehyde solution	100 mL
Sodium chloride	9 g
Tap water	900 mL

Alcoholic formaldehyde:

40% formaldehyde solution	100 mL
95% ethanol	900 mL

Bouin's fluid:

Picric acid (saturated, aqueous) 75 mL
40% formaldehyde 25 mL
Glacial acetic acid 5 mL

Martius scarlet, blue staining: follow local laboratory standard operating procedure for materials and methods.

7.5.2.2 Method

1. Fix tissue in formol saline, alcoholic formaldehyde and Bouin's fluid for 24 hours.

2. Paraffin embed and section.

3. Stain dewaxed and hydrated tissue section using the Martius Scarlet Blue method.

4. Dehydrate the section by taking it through an alcohol series, clear in xylene.

5. Mount the slide and view under a microscope.

7.5.2.3 Results and analysis

The cell and tissue components can be identified by the characteristic staining of nuclei (blue), fibrin (red/orange), red blood cells (yellow) and connective tissue (green). To compare and report the staining results, randomly select five fields of view using a ×10– ×40 objective and assess the intensity of staining of selected section components and integrity of cell and tissue structures. Pay particular attention to the degree of cellular and structural shrinkage and chromatin clumping; evaluate any shift in colour balance in staining of connective tissue and cell components (Figure 7.3). Consider the effects observed in relation to the effects of the different fixatives on the permeability of the cell and tissue components.

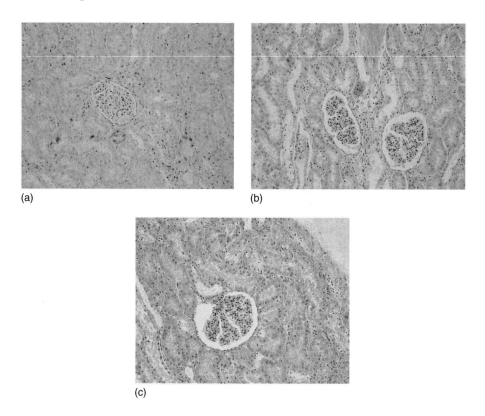

(a)   (b)

(c)

Figure 7.3 Light micrographs of MSB-stained porcine kidney sections (×20): (a) tissue fixed in formol saline; (b) tissue fixed in alcoholic formaldehyde; (c) tissue fixed in Bouin's fluid

Suggested references

Baker, F. J., Silverton, R. E., and Pallister, C. J. (1998). *Introduction to Medical Laboratory Technology*, 7th edn. Oxford: Butterworth Heinemann.

Cook, D. J. (2006). *Cellular Pathology: Introduction to Techniques and Applications*, 2nd edn. Bloxham, Oxfordshire: Scion.

Gowenlock, A. H. (1988). *Varley's Practical Clinical Biochemistry*, 6th edn. Melksham, Wiltshire: Heinemann.

Mitchell Lewis, S., Bain, B. J., and Bates, I. (2006). *Dacie and Lewis Practical Haematology*, 10th edn. Oxford: Churchill Livingstone.

Reed, R., Holmes, D., Weyers, J., and Jones, A. (2007). *Practical Skills in Biomolecular Sciences*, 3rd edn. Harlow: Pearson Education.

8

Development of knowledge and competency for biomedical scientists

8.1 Introduction

Previous chapters have shown that acquiring scientific knowledge, developing technical expertise and adopting a professional attitude are all essential and complementary elements of working as a biomedical scientist. Providing evidence for these skills, knowledge and competencies is a fundamental requirement for registration with the Health Professions Council (HPC). This is usually achieved by gathering a significant and organized collection or 'portfolio' of evidence, which is assessed by work-based trainers and academic tutors and verified by an independent registered practitioner. Once registered, biomedical scientists must also provide evidence on a regular basis to show that they continue to meet the standards required, including knowledge of important facts, competency in key tasks and behaviour as professional members of the healthcare team. This chapter gives an overview of strategies that can be used for gathering appropriate evidence by trainee biomedical scientists to meet registration requirements. These can also be applied throughout the professional career of a biomedical scientist as a part of their 'continuing professional development' (CPD). This includes the use of reflective practice as a means to support and provide evidence for professional development and using knowledge of personal learning styles to enhance learning experiences. In addition to the expertise specific to biomedical science, there a number of skills required that are important to functioning as an effective employee and which are common to many professional groups. The chapter concludes with a consideration of how both profession-specific expertise and generic or transferable skills can be important in the context of employability and career progression for biomedical scientists.

An Introduction to Biomedical Science in Professional and Clinical Practice   Sarah J. Pitt and James M. Cunningham
© 2009 John Wiley & Sons, Ltd

8.2 Gathering evidence of knowledge and competency for HPC registration

Regardless of the route to HPC registration, training for biomedical scientists is often very carefully structured, involving lectures and seminars, alongside practical teaching sessions. It should also include regular assessment, so that both the trainer and the trainee can judge how much progress has been made. As undergraduates, biomedical science students learn academic skills, such as how to conduct experiments, evaluate scientific evidence, appraise published literature and write up findings according to scientific conventions. Trainees in the laboratory (as part of an undergraduate programme or while in postgraduate training) are taught how to perform particular assays and learn how to behave as professionals and interact appropriately with colleagues. An effective way to show that this training has been sufficient and successful is to gather together a portfolio of evidence. For biomedical scientists, this should be structured in a way that clearly maps the evidence against the Standards of Proficiency required by the HPC, as is exemplified by the Institute of Biomedical Science Registration Portfolio for the Certificate of Competence. The knowledge and competence required to meet each of the HPC Standards of Proficiency for Biomedical Scientists (see Chapter 1) has been clearly set out in the Registration Portfolio and evidence must be demonstrated for each standard. Box 8.1 gives a list of possible evidence which trainees could use, although it is not exhaustive and students are encouraged to be creative in the way they demonstrate their suitability to be registered practitioners. The information and suggested exercises after each chapter in this book have been designed to provide ideas for evidence for various parts of the Registration Portfolio. Since all the standards relate to all aspects of day-to-day practice, it is possible (and indeed expected) that one piece of evidence could cover knowledge or competence in more than one standard.

Box 8.1   Examples of evidence to demonstrate knowledge and competency

Annotated documents: Documents used within the laboratory (such as SOPs or policy statements) which have been anonymized and annotated to point out that the trainee has noticed the key features within the document.

Written assignments: Pieces of written work set by the Training Officer or university tutor showing understanding of the important facts on a particular topic.

Witness statements: Written accounts from colleagues that they have observed the biomedical scientist in question perform a particular task or show a professional attitude in a certain situation. Each one should be signed and dated both by the observer and the person who was being observed.

Audits: Protocol used to conduct an audit of an area of laboratory practice (e.g. safety, quality) and a presentation of the findings.

Case studies: An account of how pathology test results helped to support the initial diagnosis, monitoring of treatment of a particular patient, using anonymized copies of any results and a clear discussion of the implications of each. The case could be concerned with findings from a single pathology discipline or could be multidisciplinary.

Visual projects: A video or a series of photographs showing the trainee taking part in a particular aspect of work in the laboratory or how their skill in a certain area has improved during their training period.

Notes: Handouts (preferably annotated) from relevant tutorials, seminars or lectures which the trainee has attended and copies of slides or overhead projection transparencies used in oral presentations give by the trainee.

Minutes: From meetings at which the trainee was present.

Reflection: Explanation of how the trainee's perception and understanding of laboratory work and professional areas have changed during the period of training. This could include extracts from their reflective diary, written accounts of thoughts and feelings in response to a particular incident in the laboratory, in addition to a description of meetings for which minutes have been included and response to feedback given on assignments.

Once the trainee has completed some work and wants to present it as evidence that they have the required knowledge or are competent in one (or more) of the standards, they must show it to their tutor or Training Officer. If that person is happy with the appropriateness and quality of the evidence, then that standard can be 'signed off' to indicate that it has been satisfactorily met. When all the standards have been signed off in the portfolio, indicating that the Training Officer and other colleagues deem the trainee to have met the requirement necessary for HPC registration and to be safe in the laboratory, then an external person, who has been trained by the IBMS, will be asked to verify this professional judgement. The 'external verifier' talks to the trainee and looks at their evidence, but also assesses the quality of the training they have received and sends a report back to the IBMS on both. Thus they are able to give an objective opinion on the overall experience for the trainee and trainers. In this way, the quality of training programmes can be maintained, as well as ensuring that only suitably committed professional biomedical scientists are eligible to enter the HPC register. The Certificate of Competence which is awarded by the IBMS can then be presented to the HPC to prove that the trainee has met the required knowledge and competency for all the Standards of Proficiency. It also allows the holder to apply to become a Licentiate member of the IBMS and then start the training towards the Specialist Diploma. In some HPC-approved degrees, the verification of the trainee's competence is done throughout the undergraduate programme, so the arrangements may be different from those described here. The end result should be the same, that is, eligibility to apply to enter the HPC register. Admission on to the HPC register is not automatic, but is subject to health and criminal record checks and references.

8.3 Continuing professional development

The end result of a successful training programme leading to HPC registration should be a well-rounded registered practitioner, who is a trustworthy member of the laboratory team, but has insight into their scope of practice and limitations. Scientific understanding of the causes of medical conditions is continually deepening, which means that tests are needed for an ever-increasing range of analytes. Technical advances mean that new testing methods are regularly introduced, either to enhance or replace assays. The National Health Service is subject to influence from politicians and wider trends in management and business, so the professional environment for biomedical scientists is also constantly changing. Therefore, one of the most important skills that newly registered practitioner biomedical scientists must have is the ability to keep learning! In 2005, the HPC made it mandatory for all registrants to participate in continuing professional development (CPD), as a condition of remaining on the register. When practitioners re-register every 2 years, they are required to sign a form confirming that they have been doing CPD activities during that time. Although most of the professional bodies representing HPC registrant professions run their own CPD schemes, the HPC do not require or even recommend individuals to join them. They accept that a formal programme does not always fulfil a person's training and development needs or fit around their working pattern. However, they have published a set of standards for CPD, which state that registrants must:

- maintain a continuous, up-to-date and accurate record of their CPD activities;
- demonstrate that their CPD activities are a mixture of learning activities relevant to current or future practice;

- seek to ensure that their CPD has contributed to the quality of their practice and service delivery;

- seek to ensure that their CPD benefits the users;

- present a written profile containing evidence of their CPD upon request (www.hpc-uk.org).

The HPC have produced guidelines about which activities could be counted as CPD (shown in Box 8.2), and registrants are expected to undertake a selection of them and, crucially, record their participation. Each person has to decide for themselves what CPD they should do and how much time they should spend on it, to fit their current job.

Box 8.2   Categories of continuing professional development according to the Health Professions Council guidelines

The Health Professions Council lists five categories of CPD:

1. Work-based learning: e.g. audit, critical incident analysis, user feedback, using reflection, attending a journal club.
   These are activities which biomedical scientists would undertake as part of their normal working day, to make sure they do their best at all times and to help improve the service to patients.

2. Professional activity: e.g. teaching, being active in a special interest group, presentation at conferences.
   Biomedical scientists committed to their profession would automatically seek opportunities to do some of these – according to ability and interest.

3. Formal education: e.g. studying for a qualification (academic, IBMS diploma), doing a piece of research, being involved in the planning and development of a course.
   At various stages in their career, a biomedical scientist will take a formal course of study, as necessary, to take on an extended role or gain promotion.

4. Self-directed learning: e.g. reading a book or journal article or using a website, newspaper or TV to update knowledge.
   The idea is to then record what you have learned, by writing a reflective piece, citing your sources. It should be noted that while the HPC recommend the use of non-specialist sources of information, they must be approached critically and used wisely!

Box 8.2   (Continued)

5. Other activities: e.g. public service, voluntary work.
   Using your professional skills and knowledge outside of the usual work place can be challenging, but is good for confidence and personal development.

8.4 Professional body support for CPD

Soon after HPC registration, biomedical scientist practitioners are likely to embark on another period of structured learning in the pathology discipline in which they have chosen to specialize. This should include gathering evidence for the IBMS Specialist Diploma Portfolio; lectures and seminars might be arranged in the local area to help with the specialist knowledge required for this qualification. In addition, to help practitioners to prove that they are keeping their knowledge and skills up to date, formal CPD schemes, such as the one administered by the IBMS, have been set up. This is based around activities organized by its members at local, regional or national level and credits are awarded for attendance, according to the length of time spent doing the activity (Box 8.3). The scheme is open to anyone who is interested and IBMS members can join free of charge. Although members are not obliged to be part of the IBMS programme, this scheme provides structure to CPD activities that are required to retain registration and awards credits for doing things which are part of normal professional activities. Collection of credits through the IBMS scheme is just one way of achieving the HPC requirements, but individual practitioners can organize their learning and development in any way which suits them provided that it is clear that they are making the effort to develop in their professional understanding and skills. While some of the IBMS-run events have an attendance fee (e.g. the IBMS Congress), most employers are prepared to contribute to the cost even if they cannot meet it in full. However, many events (such as lunchtime or evening meetings) are free and sometimes food is provided through sponsorship. One of the principal objections which biomedical scientists make to attending IBMS events is that they have to do so in their own time. This is a valid point, since people often operate under great pressure in the laboratory, and so need their breaks and are tired at the end of the working day. Family commitments may also preclude going to an evening meeting without plenty of notice of the date. However, it must be remembered that CPD is mandatory for HPC registrants and it is the individual practitioner's responsibility (*not* their employer's) to ensure that they have done enough to meet the requirements.

Box 8.3 Examples of activities in the IBMS CPD scheme

- Monthly lunchtime 'journal club' at local Trust

- Evening lecture arranged during a local IBMS branch meeting

- Training workshops held by equipment manufacturers

- Attendance at IBMS Congress

- Organizing or presentation at CPD activities

- Writing a relevant article for publication

- Active membership of local IBMS branch or national committees

- Formal education programmes e.g. degrees, short courses

- Teaching students and trainees at a university or through laboratory-based training

- Submitting a written reflection on a CPD activity

Participants in the IBMS scheme can collect certificates of their attendance at accredited activities, record training in which they have been involved (either as the trainee or the trainer) and write reflective accounts of their learning and development in the form of a structured portfolio. At the time of writing, the IBMS are about to introduce an 'on-line system' for members to record their CPD activities and reflections. Further details of the scheme can be found at www.ibms.org/professional.

Experienced practitioners can also use their CPD to take advantage of the IBMS's participation in the Chartered Scientist scheme. A number of professional bodies for scientists are involved in this and there is a central register of practitioners who have been accepted as meeting the necessary criteria. It is intended to give recognition to individual scientists of their specialist expertise and also to improve public trust and confidence in scientists – since people are familiar with the concept of 'chartered' professionals' (e.g. chartered accountant, chartered engineer). Biomedical scientists who have been IBMS members at a corporate grade (Member or Fellow) for at least 4 years, hold a Master's-level qualification and have evidence that they are actively undertaking CPD, are eligible to apply. Being on the register of a Chartered Scientist scheme can benefit biomedical scientists, as it allows them to demonstrate to colleagues, employers and people that they interact with outside of the laboratory that they have a recognized professional status. Chartered Scientists use the designation CSci after their name.

8.5 Reflective practice for biomedical scientists

Reflection is something which good scientists and professionals do to help deepen their understanding of topics or improve their practice. It is a way of thinking analytically about one's understanding of a scientific concept or competence in practical skills and using the outcomes to deepen knowledge and improve practice. In the professional context, teams of people who work together can also use reflective practice to change their working patterns and thus provide a better service.

If you spend the journey home after work thinking about what happened during the day and how any problems which occurred might be avoided in future, then that is reflection. Mulling over the information given out during a lecture and seeing how this fits in with what you already knew are reflection. Taking seriously any feedback which you have been given about written work, practical skills or behaviour (be it good or bad) and deciding how you can change things for the better are reflection. When biomedical scientists think about feedback from service users and discuss how to use those comments to enhance the service, they are reflecting. In common with all healthcare professionals, biomedical scientists are now required to translate these reflective thoughts into writing, which many people find hard if they are not used to doing it. All biomedical scientists, however senior and however long they have been registered, are now finding that they must train themselves to incorporate 'reflective' practice into their professional lives, as part of continuing professional development (see above). There are three main areas that biomedical scientists need to learn to reflect on:

1. Their individual development as a scientist and a professional.
   By thinking about what they already understand already and what more they need to know, practitioners can identify what they are really good at and interested in – and also where they could benefit from more learning or training. It should also help to highlight how they learn best – i.e. their 'learning style'.

2. The quality of the service that their department is providing.
   Biomedical scientists and their colleagues in pathology usually work on the assumption that service users are happy with what their department is doing. However, if there is a mistake or a complaint, the department needs to learn from it. One way of learning is through 'critical incident analysis', which is a tool to reflect on exactly what happened, which can help to decide on ways of preventing that mistake or complaint again. As a professional, it is very important not to blame any individual people for what happened; the whole team or department has to contribute to the reflective process in some way and take responsibility for the failing and for putting things right.

3. The scientific and technical standards of work in their laboratory.
Biomedical scientists keep up to date with scientific developments and advances in technology for diagnostic tests by going to conferences and external lectures, reading journal articles and talking to colleagues who work in other departments or hospitals. This information can then be fed back to the rest of the team and can be used to help in making decisions to change test methods or upgrade machines.

8.6 Approaches to reflective practice

It is a good idea to give your reflections some structure and to take advice from academic or work-based tutors about effective ways to do this; however, the style will be personal and possibly unique, so there is no 'right' or 'wrong' way of doing it. It is advisable to keep a reflective diary, but it does not matter whether you write up your thoughts by hand in a large notebook, type them up on a computer or make an audio diary. It should include structured accounts of your experiences and insights recorded regularly (for example, daily or weekly) and may also contain discussions of particular achievements or analyses of problems as they arise. The format is less important than the outcome and the only time to consider changing how you document your reflections is when the system that you are using is not working for you. By keeping a record of how you are thinking at a particular time, you can plan future learning and also demonstrate how your understanding is deepening and your skills improving over time. Elements of reflective practice are increasingly being incorporated into academic programmes, so students may be asked to reflect as part of a piece of coursework – for example, as part of a practical write-up. Universities throughout the UK have introduced Personal Development Profiles (PDPs) for students in recent years. These are designed to help the student document their learning and development, so that they can identify their individual strengths and areas which need improvement. Constructive reflection is a key part of this process and, once acquired, it is a skill which should help graduates in their future careers (see Section 8.9). Similarly, the IBMS Registration Portfolio (and indeed Specialist Portfolio) guide the trainee to reflect on their progress periodically (for example, on completion of each section). Thus, in the workplace, trainees are often given worksheets to complete after a training session which include reflection. It is important always to bear in mind that *reflecting* on what you have been doing is not the same as simply listing activities; reflections must show how the person's understanding of theoretical topics and confidence in practical skills are developing.

Although not always possible, where a student is engaging with a discrete or well-defined activity or experience, either as a single event or over a period of time, it can be very helpful at the outset to set out some questions that can be used as a reference point for reflective diary entries and evaluation. Box 8.4 gives a list of suggested structured questions which can be used for the daily or weekly entries into a reflective diary.

Box 8.4   Structured questions for daily or weekly reflective diary entries

The following series of structured questions can be used each time you reflect on an activity or experience:

Where did I start from?

1. What was the aim of the exercise learning session, course or experience?

2. What do I know about this topic/practical technique already? (e.g. refer to previous lectures attended or practical experience).

3. Are there any specific aspects that I expect to encounter difficulty with or require additional attention?

Where am I now?

4. What new theoretical information or practical skills have I learned?

5. How has what I have learned changed my understanding of this topic or practical task?

6. Have any unexpected new experiences or insights been gained that are relevant to this or other aspects of my professional or academic development?

7. Has my overall aim been met and if not is there anything more that I need to learn to fully understand the theory or be fully competent in the practical skill?

Each time you are ready to write in your diary, go back and read your previous entry, to remind you what you had written and what you were thinking about when you wrote it. Try and build on your thoughts and ideas as you go along. Also, keep checking that what you have written would show another person *how* you have been learning and how your understanding is deepening – not simply what you have been doing.

As an example of how these questions could be used, consider a trainee biomedical scientist learning about gel electrophoresis to separate serum lactate dehydrogenase (LDH) isozymes in a practical session. As a starting point, the first three questions ('where did I start from') help establish a reference points to guide the reflective process. The answer to the first question posed in Box 8.4 ('what was the aim ...?') could be 'to become competent in the use gel electrophoresis to separate serum proteins'. The trainee might have heard or read about gel electrophoresis but never attempted to do it themselves; so for question 2 ('what do I know ...?') they could write 'I was confident that I understood the theory of gel electrophoresis but I had never used the equipment before'. For question 3, concerning possible difficulties anticipated, a student might state 'I am unsure about how well I can manage pipetting of very small volumes to load gels'. At the end of the learning session or experience, or if a longer course at appropriate intervals, the student can use questions 4–7 ('where am I now?') to guide their diary entries and overall refection. Responding to questions 4 and 5 concerning new knowledge and understanding should help a student appreciate how much they have gained in these aspects and whether they would feel confident about undertaking the task again. Answering question 6, related to unexpected new insights or experiences, could help a student appreciate the full range of new skills and knowledge gained not anticipated at the outset; for example, they might have been asked to make up buffers from stock solutions and learned how to use new equations to calculate appropriate weights and volumes. There may well be aspects of the technique encountered that the trainee is still unsure of, such as how to interpret the results from the gel electrophoresis and the clinical significance of LDH enzymes. This could be included in the response to question 7 usefully to highlight areas for further study or development or be a basis for the next step in their learning about the technique – for example, by discussing the meaning of the results with their Training Officer. Overall, looking back on these answers before making a subsequent diary entry on this subject should help trainees to see that their competence and confidence with gel electrophoresis in the context of clinical biochemistry have improved. It can be useful to ask a tutor or senior colleague to read (or listen to, if you have an audio diary) your reflections, particularly your first attempts, to make sure that you are using a style which not only works for you, but is truly reflective.

8.7 Critical incident analysis

Another way of using reflection is to look in depth at a particular event, which stands out as significant for some reason (a 'critical incident'). It can be done individually (for example, when you finally understand a concept that you have been struggling with) or in groups (for example, staff working in a particular section of the laboratory after feedback from a user of the service). In the context of reflective practice, a range of occurrences could be chosen as critical incidents and they should include both positive and negative experiences. Examples include:

- When something unusual or unexpected happens (e.g. a friend turning up at the laboratory, without prior arrangement, during the busiest part of the working day).

- When something goes wrong (e.g. a mistake with patient results, breach of safety, misuse of expensive reagents).

- When something really good occurs or the work goes unexpectedly well (e.g. an urgent result processed in a fast time).

- An ordinary, everyday event which went as expected, but you learned a lot from it (e.g. you mastered a particular technique).

It is very important to remember that the 'incident' is probably the end result of a series of events. In the example of a student finally understanding a particularly difficult scientific concept, it is because they have realized the relevance of a crucial piece of information. However, that student must have been gathering knowledge and thinking about that concept for some time already in order to fit the last vital fact into the jigsaw. Similarly, when an accident or error occurs in the clinical laboratory, it is rarely the sole responsibility of one person – their colleagues and managers have often made mistakes which contributed to the problem. Thus, the key to effective critical incident analysis is an open mind. To analyse an event, it is first necessary to describe what happened precisely and objectively and identify why the occurrence is of interest (to you as an individual or to the laboratory team); then one must think about what the people involved might have said or done to contribute to the actual incident which was observed. As for a reflective diary, it is best to use a structured approach to this technique, and Box 8.5 outlines the five key steps to effective critical incident analysis. Note that truly effective use of this type of reflection includes step 5, dissemination of findings to the appropriate audience.

Box 8.5   Key steps in reflective analysis of a 'critical incident'

Step 1: Description of the incident
Write a clear account of:

- What happened

- How it happened

- When it happened

- Where it happened

- What was going on in the general area at the time.

Step 2: Identification of possible causes of the incident

Think carefully about why the incident happened. Write a logical (chronological, if appropriate) description of the events which led to the incident

Step 3: Reflection on the incident

- What things have been highlighted that you were not previously aware of?

- What things had you known about but not fully understood before this incident?

- How has your thinking on this topic been changed?

Step 4: Identification of changes to implement

Use your reflections to think of how laboratory practice or individual behaviour should be changed as a result of analysing this critical incident. For example: change in laboratory protocol/SOP for particular procedure; introduce more staff meetings; have more opportunities for staff training; buy a new text book.

Step 5: Dissemination of the outcome of the analysis

Select an appropriate way to discuss the insights gained from the analysis with colleagues, for example during a laboratory staff meeting or during a university tutorial.

8.8 Learning styles

It is clear that biomedical scientists must continue to learn new topics and skills throughout their career, but it is important to realize that different people learn in different ways. For example, when learning a new assay, some people prefer to start with abstract scientific concepts and feel confident about those, before embarking on learning the practical skills; other people learn best by trying the assay and using the tangible results to help them understand the theory. The way in which a person prefers to approach new information or acquiring skills is called their 'learning style'. It is useful to understand one's learning style for two main reasons, the most obvious one being so that you can identify the ways in which you learn most effectively. The second is to appreciate that sometimes when you struggle to understand something it may be the way the information has been presented which is the barrier, not the concepts *per se*. Thus, knowing how to best adjust your approach to the task to suit your learning style may be a significant help to the learning process.

When researchers started to investigate how people learn, they found that when presented with a task or problem an individual usually follows a series of discrete stages towards understanding. One of the best known models was developed by Kolb, who described four steps that take place during learning, described as Concrete Experience (where a learner engages in a new experience), Reflective Observation (analysis of experience), Abstract Conceptualization (processing and generation of related ideas and theories) and Active Experimentation (planning and undertaking practical action to test ideas and theories). When an individual has seriously engaged with the matter of trying to learn about a subject, they will go through all these steps in turn, possibly several times – the so-called 'Learning Cycle'. However, different people tend to start at different places in the Learning Cycle. For example, when presented with a new task or problem some prefer to 'dive in' directly to tackle a problem and in part through trial and error find this approach to be the most stimulating. Others learn best by some preliminary observation and research into the task so that possible pitfalls can be identified before any practical engagement. Another individual might prefer to read the SOP or diagnostic assay kit instructions first and to make sure they are fully happy with the principles underlying the task before attempting anything practical. Another learning style might be characterized by individuals who prefer to have the theory and method explained so they can plan the method before attempting under supervision. A system commonly used to describe this variation in learning styles was proposed by Honey and Mumford (1992), which categorizes people into 'Activists', 'Reflectors', 'Theorists' and 'Pragmatists'. Again, their strengths are best applied at different stages of a Learning Cycle, which corresponds closely to Kolb's model (Figure 8.1). In practice, most people are a mixture of the four types and usually vary their approach slightly depending on the situation. For example, a biomedical scientist whose learning style can be best described as a 'Theorist' with strengths in dealing with abstract concepts would be more likely to approach learning how to enter data on a new computer system by doing the task (i.e. engaging as an 'Activist' in a 'concrete experience') rather than to decide to read up about the underlying computer programme used to handle the information.

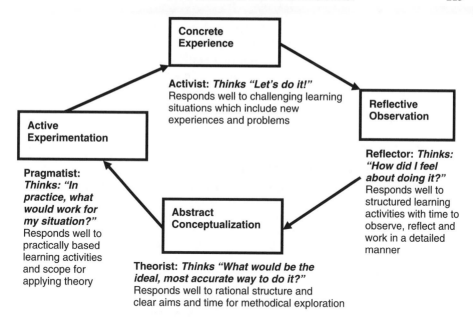

Figure 8.1 The learning cycle and learning styles [adapted from Kolb (1984) and Honey and Mumford (1992)]

Understanding the way in which learning takes place can not only help tutors in the design of teaching and learning programmes, but also guide learners in recognizing their strengths and weaknesses in different abilities that are used during the learning process and whether they might benefit in modifying their approach. It can be enlightening to understand your individual learning style and relatively simple methods to determine your own profile can be followed (e.g. see Kolb 1984; Honey and Mumford 2006; Jasper 2003)'.

8.9 Planning a career as a biomedical scientist; generic skills and employability

Acquiring a professional qualification can provide eligibility to a particular career pathway or post. However, when employers are selecting the best candidate from a number of applicants with the same or similar entry qualifications, there is a diverse portfolio of skills and attributes that need to be considered, including 'transferable' or 'generic' skills such as communication, team working and time management. Work placements in higher education courses allow students to develop vocational and professional skills that can only be acquired in the workplace. In addition to these profession-specific attributes, the workplace experience can also offer excellent opportunities for students to develop and enhance a range of generic skills that are valued by employers and an applicant can improve their chances of employment by being able to demonstrate convincingly all the relevant skills, understanding and personal attributes that are good indicators for success in employment. Box 8.6 lists a selection of transferable and generic skills looked for in recent job descriptions for biomedical scientist posts at career entry and after a few years' experience.

Box 8.6   Some generic and transferable skills used as selection criteria by employers in job advertisements for biomedical scientists

Early career (Band 5/6)	Later career (e.g. Band 7/8)
Shows initiative	Willingness to show initiative
Good written and verbal communication skills	Good communicator and interpersonal skills
Ability to communicate within the team to deliver an effective service	Ability to work as part of a close team and encourage good team performance
Proven ability to work independently	Be able to work effectively as an individual
Ability to prioritise and organise the work of yourself and others	Systematic approach to work
Time management skills	Ability to work under pressure
Ability to train other staff	Ability to teach and train others
Willing to learn	Provide good professional leadership skills for all staff grades
	Efficiently organise the day-to-day requirements of the department

---

Although there are some requirements that are clearly dependent on a person's experience in the 'later career' specifications (such as organizing the department's needs and leadership skills), there is considerable overlap and similarity in the nature of the attributes required at both grades. This emphasizes the importance of not only developing these attributes as a student but also maintaining them throughout a career. When assessing candidates, employers look very carefully for evidence of how an applicant meets the criteria set for a particular position and this can be provided by the applicant through the written application, the interview process and referees' statements. Some criteria are more easily evidenced than others. For example, experience in working with a set of particular techniques can be demonstrated through giving appropriate work experience details and answering related interview questions. However, it can be more difficult to provide evidence for many of the generic skills commonly looked for. Reflective practice is increasingly recognized as an enhancer of good quality academic and lifelong learning achievements, including those which underpin employability. Being aware of the skills important to professional practice and working as a reflective practitioner can also be a very effective means of compiling a record of appropriate evidence which can be used in preparing written applications or interview preparation. Some of the more difficult (but not unusual) interview questions, such as 'give an example of how you have performed well under pressure', may then be prepared for by a review of appropriate experiences gained and recorded as part of reflective practice.

8.10 Conclusion

This is an exciting time to begin a career as a biomedical scientist. As this book has shown, biomedical scientists can work in a variety of disciplines within pathology and can acquire a range of professional skills. The nature of the job has changed dramatically over the past 20 years due to rapid technological advancements, increased automation and the use of computerized support for all aspects of the laboratory work. This pace of change is unlikely to slow down in the foreseeable future and there will be opportunities for biomedical scientists to take on challenging roles as healthcare practitioners, scientists, managers and leaders. The roles of all healthcare workers are continually changing and, in future, chances to become involved in patient care in hitherto unexpected ways (such as within primary care) are likely to present themselves to biomedical scientists in future. An individual practitioner taking up a post as a biomedical scientist in the 21st century can expect a long and varied career if they would like one, but they must be prepared to be flexible, to participate fully in CPD and to engage with other members of the profession at local, national and even international level.

Quick quiz

1. What does the abbreviation CPD stand for?

2. List five types of evidence which would be acceptable to show competency in the IBMS Registration Portfolio.

3. List three activities which are suggested by the Health Professions Council as suitable CPD for registrants.

4. What stage in the Learning Cycle comes after Concrete Experience (Activist)?

5. What are the three main areas that biomedical scientists need to reflect on in their professional practice?

6. Why is it important to keep a reflective diary?

7. Give one example of a 'critical incident'.

8. What is the first step of the critical incident analysis process?

9. Name two 'generic' skills which a biomedical sciences graduate might have.

10. Give two key differences in role and responsibility between a practitioner at Band 5 and a senior specialist practitioner at Band 7.

Suggested exercises

1. Make a list of the continuing professional development activities that are open to you as a student or trainee biomedical scientist and plan to attend at least one event in the next 6 months.

2. Attend a seminar or journal club and use structured reflection to identify clearly what you learned and highlight any related areas which you could investigate further.

3. Ask your Training Officer or another senior colleague to help you identify an adverse event or incident in the laboratory which would be suitable for critical incident analysis. Apply the 'critical incident analysis' technique to this event and write up your findings in a format suitable for the staff notice board within your department.

Suggested references

HPC (2006). *Your Guide to Our Standards for Continuing Professional Development.* London: Health Professions Council (www.hpc-uk.org).

Jasper, M. (2003). *Beginning Reflective Practice.* Cheltenham: Nelson Thornes.

Honey, P., and Mumford, A. (1992). *The Manual of Learning Styles.* Maidenhead: Peter Honey Publications.

Honey, P., and Mumford, A. (2006) *Learning Styles Questionnaire.* Maidenhead: Peter Honey Publications. See also www.peterhoney.com for further information and resources.

Kolb D. A. (1984). *Experiential Learning: Experiences as the Source of Learning and Development.* Englewood Cliffs, NJ: Prentice Hall.

Reed, R., Holmes, D., Weyers, J., and Jones, A. (2007). *Practical Skills in Biomolecular Sciences*, 3rd edn. Harlow: Pearson Education.

www.ibms.org/education, for information about the Specialist Diploma and further IBMS qualifications.

www.ibms.org/professional, for information on CPD and other areas of professional interest.

Index

Note: References to figures are in *italics*, to tables in **bold**